Providence Lost

Providence Lost

A Critique of Darwinism

Richard Spilsbury

LONDON

Oxford University Press

NEW YORK   TORONTO

1974

Oxford University Press, Ely House, London W1

GLASGOW NEW YORK TORONTO MELBOURNE WELLINGTON
CAPE TOWN IBADAN NAIROBI DAR ES SALAAM LUSAKA ADDIS ABABA
DELHI BOMBAY CALCUTTA MADRAS KARACHI LAHORE DACCA
KUALA LUMPUR SINGAPORE HONG KONG TOKYO

ISBN 0 19 212963 5

*Printed in Great Britain
by Ebenezer Baylis and Son Limited
The Trinity Press, Worcester, and London*

Contents

Preface

I have two main aims in this book, one explicit, the other implicit. The first is to explore some limitations in our scientific thinking about man, with special emphasis on the Neo-Darwinian concept of human evolution. This theory seems to me inadequate to account for the uniqueness of man, as exemplified by his acute self-awareness and world-awareness, his capacity for cultural development, his linguistic and creative powers, and his sense of values.

To bring out this inadequacy, I have reformulated and sharpened certain perennial objections to Neo-Darwinism and have added some new ones. I have laid bare underlying assumptions on the acceptance of which the plausibility of the theory depends. I believe the full understanding of evolution lies beyond our present horizons. Neo-Darwinism seems in danger of replacing one set of dogmas about the origin of man by another set. We can have no guarantee, or right to expect, that this amazingly productive process will be fully comprehensible to our minds that are its product.

My second, implicit aim arises from the fact that I who write am a philosopher. I shall not attempt to define this word and have no wish to make sharp the boundary between philosophy and non-philosophy; on the contrary, I believe the narrowness of most academic philosophy (and there is, alas, little unacademic philosophy of any value being produced) derives from this demarcation-urge. I hope to show in this book that it is possible for philosophers to make a critical and constructive contribution to questions of natural philosophy that have the deepest relevance for our world view. No one should have to take this ready made, or at second hand, from the Authorities of the age.

Fundamental questions about the nature and origin of man are no respecters of academic compartments. This great theme of evolution can be surveyed from many viewpoints, besides those of Palaeontology, orthodox Genetics, and molecular Biology. Some of the more imaginative applications of the latter run the risk of creating a molecular mythology.

A fashion of philosophizing which turns away from such questions as the ultimate randomness or purposiveness of life and death, seems to be heading for extinction as a subject of general concern and non-specialized significance. It is probably true that in the past Philosophy has generally lain on the very edge of human capacities, testing them cruelly. Nevertheless its present poverty is a serious loss for the whole of our shared intellectual and imaginative life, and does little to check the drift towards the more banal, newspaperish aspects of our culture. Philosophy, which originated in wonder, should remain its guardian.

A useful service that philosophers can perform is to persist in questioning the prevailing beliefs of their age, and especially those beliefs about the nature and origin of man that are unlikely to be decisively settled by observation, experiment, or any other means. If everyone came to agree on these essentially contentious issues, real independence of judgement would be impugned.

I am very grateful to Mr. Arthur Koestler, Mr. Derek Parfit, and Professor C. H. Waddington for their reading of an earlier draft. They do not, of course, endorse all my heresies.

The permission of William Collins Sons & Co Ltd., London, and of Alfred A. Knopf, Inc., New York, is gratefully acknowledged for the extracts quoted from Jacques Monod's *Chance and Necessity* in the translation of Austryn Wainhouse (copyright © 1971 Alfred A. Knopf, Inc.)

1 Evolution

The tedium, the irrelevance, the archaism of investigations of human origins. Palæontology is self-evidently a dry-as-dust pursuit. Let the fossilized dig for the dead. The thought of those pre-human Ages, vast, silent, and seemingly purposeless, is more likely to bore than over-awe one. One may shrug them off with indifference, with the cosmic impiety of a yawn. One of the unappreciated gifts of scientific naturalism is that it removes the bad conscience from boredom before the out-dated works of nature. Why should we pretend to be excited or awed by the messages from a remote earthly past that we prise from rocks, any more than by the signals from the deep cosmic past that we can effortlessly pick up by looking at the fossilized night sky, any of whose luminaries might be long extinct before its light reaches our eyes?

Such rude and crude thoughts might occur to anyone convinced that the evolutionary past is fortunately *passé*, and largely irrelevant to men's present endeavours and enterprises. Given the belief in a large dose of fortuity in the origins of man, and of everything else, such a person may conclude that any value or significance in human life has to be created by men, despite or even in defiance of the purposeless processes of organic evolution. It is no good looking to this value-barren past for inspiration or guidance. Men can have nothing to learn from a process that is essentially sub-human, undirected, and unmotivated. Even to study such a process in great depth and detail, in a detached scientific spirit, may seem a waste of time and talent that would be better employed in the study and promotion of human creative processes and productions. The scientific 'running down' of the universe, this 'depreciation' of life as an interplay of 'chance and necessity', may at length be a discouraging influence on science itself. For how long can the mindless and the meaningless be expected to attract the enthusiastic attention of intelligent and creative individuals? Is a scientific vocation a kind of banishment from the realms of mind and value?

Whatever answers one may give these questions, the sense of discontinuity and dissociation between human ends and values, and the evolutionary process that brought man into being, is a critical aspect of our present mental epoch. The difficulties and incoherences implicit in this divorce between man and his roots will be one of the themes I shall concentrate on in the following chapters.

The question how Homo Sapiens came into existence has, for many men, the same kind of compelling interest that the question of their individual origin has for most children. Our interest in man's ancestry resumes and extends our first questions about ontogeny, as each backward step invites a further one into the depths of the past. The child's belief that he had a beginning generally precedes his discovery of its nature, and, historically speaking, the belief that the human race had an origin has yielded different ideas about its nature. For the greater part of human history, this belief was held without reference to evolutionary notions or the evidence of fossils. To a certain limited extent, the advancement of science consists in finding stronger reasons for old beliefs. The rudiments of evolutionary thinking are to be found in the poem by Lucretius and his Greek sources.

Man's interest in his origins, individual or racial, has something in common with his concern over endings, individual or racial. These are related events, for men that are aware of existing temporally and temporarily. But as life-brackets, they differ in their implications of dependence. Man cannot originate himself as an organism or as a species. He depends on other pre-existents. But he may be self-extinguishing, without external aid. A solipsist can accept his coming death, whether willed or unwilled, but not his birth and inheritance, so far as these imply that his powers of mind and forms of consciousness have come to him from other beings like himself. Descartes was no solipsist with regard to other human beings, but he was a 'species solipsist', believing animals to be machines, and could not have accepted the idea that man's mental powers had an evolutionary history. One gap he especially emphasized was that between human and animal communication, and his distinguishing criterion, that of free invention versus stereotyped repetition, is one that is still used today, with some qualifications.

If any question about man may be described as fundamental, the question of his origin and the 'founding' of his species may be. There is no more fundamental way of probing the relation between the human and the non-human than by attempting to think man out of existence and into existence, and by confronting the transition from an unmanned to a manned world. All other transactions between man and world presuppose this unique, original, and originative relation. Even if one believes that the proper study of mankind is man, he cannot possibly be understood in isolation. In many ways his existence points beyond itself. If he has any unique qualities not wholly of his own making, these must be referred to non-human sources. Any general philosophy of man that takes his existence for granted is shirking a basic issue.

This forgetfulness that man had an origin might be excused on the grounds that this is of purely antiquarian interest, and that palæontology is a dry-as-dust delving into a past that has no significance for our lives today and in the future. But questions about man's past evolution are continuous with questions about his present dependence on his inherited constitution. These are limiting questions, in that they evoke one kind of limitation on man's self-sufficiency and independence. They transport us out of ourselves, in an indebtedness we find exceedingly difficult to clarify in a decisive way.

By 'dependence', something more than mere derivation is meant. We *rely*, within limits, on the innate 'wisdom of the body' in its self-regulating processes, and its power of fighting germs and injuries, to keep us healthy, and we rely on our sense equipment and mental capacities in our practical and theoretical grapplings with the world and our pursuit of an intelligible order. This reliance on inherited capacities needs to be referred back to the way in which these were originally formed and developed.

This sense of present indebtedness to past endowments native to our species, and the inescapable need for reliance on these in all possible explorations of the world, have traditionally been associated with a religious attitude to life. The transition to a naturalistic conception of human evolution has left unchanged man's dependence on a past beyond his control, so far as his inherited endowment is concerned. Even if he became capable of changing this in a controlled manner, such reforms, for better or worse, could only be instituted by 'unreformed' men using their native wit.

The transition to a naturalistic concept of evolution has made it harder to express, in a meaningful way, this sense of man's dependence on processes that stand 'behind' him, outside his control.

Our lives fall apart if not informed by some sense of purpose and direction, but how can this be meaningfully related to the non-purposive processes of organic evolution, as generally conceived today? It appears incongruous for a purposive and thinking being to be dependent on the non-purposive and non-rational, with respect to his mental powers. This is a paradoxical form of dependence, and one that we are as far from understanding today as in the past.

The sense of awe that used to attend the notion of man's origin, when this was ascribed to supra-human powers, seems inappropriate when latched on to evolutionary processes, even though these may be said to have produced supra-human results, which to a great extent are beyond the present power of men to reproduce experimentally, and which should still arouse wonder. The results may be supra-human, but the 'means' appear to be sub-human, in their generally believed aimlessness. Nietzsche saw in Darwinism the embodiment of the principle of the greatest possible stupidity of Nature. Certainly there is nothing awevocative about such concepts as natural selection and gene-mutation. One may even feel rather contemptuous at such clumsy mechanisms and so crude a 'technology' of evolutionary production and invention, and feel amazed at the capacity of such processes to produce a being capable of transcending them, and of instituting a different kind of evolution, informed by conscious forethought and planning. Man is no mere evolutionary 'product', but one who participates actively in an evolutionary process which goes beyond anything previously known, and which cannot, as far as is known, exist apart from man.

Scientific thinking about man, and about human origins, tends to replace wonder at by wondering about, awe by curiosity, marvelling by puzzlement, and the apprehension of ultimate dependence by the comprehension of causal mechanisms. Wondering at and wondering about, though contrasted in attitude and orientation, do not necessarily exclude one another. Kant wondered at the starry firmament, but also wondered scientifically about it.

To wonder about man's origin is to think of this as a puzzle or problem to be cleared up, before passing on to something else. To wonder at man's origin is to experience a deep-rooted dependence on the non-human which, as the ultimate source of human understanding, may not be wholly understandable, which both limits and transcends man. This sense of the *depth* of implication in man's having originated has been greatly reduced by scientific conceptions, which also imply the isolation (or 'alienation') of man in certain crucial respects. Thus, what is sometimes called 'scientific rationalism' implies, in this context, a belief in the total absence of rationality from the evolutionary 'thrusts' or 'pressures' which brought man into being. Whether this belief in the absence of rationality should be described as 'rational', is probably not important. It is a basic decision or presupposition or postulate, capable neither of proof nor disproof, but subject to critical appraisal in the light of its theoretical successes and failures. There are no decisive logical short-cuts. It is tempting to argue that human rationality is totally discredited if it has 'evolved' non-rationally; but who can believe that the validity of Pythagoras' theorem turns on the issue of Pythagoras' descent?

A possible sceptical comment on the above: 'What's so wonderful about man's having originated? This is neither more nor less wonderful than the origin of any other species—not to mention other, far greater and more fundamental origins, such as the origin of life, or time, or the cosmos.' This may or may not be fair comment, but it misses a point. This is that man's having originated involves us all, *as men*, in a special and unique relationship with the non-human, and a unique sense of 'owing our existence' to something or some non-thing beyond ourselves. Cabbages, etc., lack this sense, and do not have to face its implications. Their derivation from non-cabbages concerns them not. Everything on the earth, perhaps in the universe, is in this respect a cabbage, save man. Man does this wondering for them, as it were—but is not involved in the same way over their having originated as over his own origins. Concern over origins originates with man, in his quest after his own. If cabbages were similarly concerned, they would be primarily concerned about themselves, no doubt. And the fact that man alone is capable of wonder at, and about, his origins, is itself cause for wonder.

Proximity to origins is in no way a necessary condition for preoccupation with them; perhaps even the reverse. To use a rather dubious analogy, new-born infants are unaware of their conception, which is, for them, inconceivable. But who can tell what ideas flashed through the minds of pre-historic men? Maybe they surmised their descent from animals, if they had a language capable of representing this.[1] Only a tiny proportion of the ideas of men gets preserved and recorded. Many discoveries are made possible by this. But in the silence of our remote forbears, we have presumably lost no vital evidence concerning our human origins. They were not privileged witnesses. Origins can only be inferred retrospectively, as extinction can only be feared prospectively.

It may be that reliable inferred knowledge of how man originated or evolved, is beyond man's reach. He cannot prise this secret from the past, unless it be somehow open to disclosure in the present. It seems unlikely that we shall ever know how language originated, with the same degree of justified confidence with which we may expect to know how a child today learns his native tongue. All the theory in the world is incapable of compensating for the absence of decisive present evidence concerning that unique origin, to which all of us men today are so deeply indebted. Have we any better reason for supposing we can know how man originated and became differentiated from his animal ancestors—even on the dubious assumption that this question can meaningfully be separated from the question of the origin of language? It is not only in respect of language that man may be believed to be unique, even though the use of language may underlie certain other forms of uniqueness, such as the making—and breaking—of contracts, the waging of war, and the negotiation of peace.

The prime assumption made by those who would infer how man originated from present evidences is that the latter provide the basis for a convincing argument by analogy from the present to the past: so that our present understanding and observation of

---

[1] Some forms of Totemism seem to contain an evolutionary germ.

how new beings originate can be applied to the analogous case of the origin of man—or any other past origin.

These analogies brought to bear on the past often originate in human activities. As a source of present change and innovation, man knows himself best. It is his 'originality' that is reflected in the history of the development of new and refined artefacts, a history that led, by analogical inference, to the concept of organisms as divine artefacts.

Another well-known analogy is drawn from a different field of human activity, selective breeding and crossing, whereby new self-perpetuating 'lines' and varieties are brought into being by human intervention in the breeding process. As with the previous analogy, this one fixes on changes brought about by man for his own ends; this weakens the analogy with natural evolutionary change, without necessarily invalidating the use made of the analogy. A great difference between the two analogical inferences— from human to divine design, from human to natural selection— is that the first goes in the direction of increasing and magnifying purpose, while the second goes on the opposite path, of subtracting and discarding purpose. The first is an inference from a lesser to a greater design, the second, from intended changes and purposive selection to the non-purposive and unintended.

The analogical derivation of the concept of natural selection may now be considered logically irrelevant: since natural selection has been shown to work in the way required by the theory, the analogy with human selection, whatever its original heuristic value, can be disregarded. Selection in biology is a concept that has been stripped of all human associations, as the concept of elective affinities in chemistry has been.

There remains, however, the question of the range of applicability of the concept. If it is to be applied to past evolutionary changes different from any that are now observable, there is still the need for analogical inference from present to past. Tentatively, it may be questioned whether the belief that natural selection has been the most important factor in evolution is wholly uninfluenced by the analogy with human selection; at least, there is a suggestive, perhaps persuasive, parallel between the notion that human selection of progenitors has been the method whereby man has 'improved' on the wild stocks of animal and plant, and that natural selection has been the means whereby 'nature' has improved

the adaptive viability of organisms, by selection of the better
adapted as progenitors. This kind of correspondence, or uni-
fication, however satisfying æsthetically, has no logical force.

The belief that evolutionary changes seen to occur by natural
selection in the course of the twentieth century provide a firm basis
for an analogical retro-inference to what has been going on ever
since the dawn of life, seems somewhat rash. It is almost as though
some Martian historians or sociologists, after a month's investi-
gation of how changes come about in human societies during that
month, should then pore over archæological records and conclude
that human history, from its earliest beginnings, has been determined by the same causal factors that they made out during that
month.[1] This rashness may be indicative of a strong will to believe
'in' a theory that chimes in so beautifully with the prevailing
irrationalistic philosophies of our time. It certainly 'fills a need'.
Its rather weak empirical basis thus becomes veiled, possibly. It is a
fine example of predominantly rationalistic and deductive
methods used in defence of irrationalism.

Even if new species were observed to originate by natural
selection, no such contemporary analogue can be expected for the
enormous increase in complexity, versatility, sensitivity, and
powers of organisms since the earliest forms, or for the cumulative
elaboration of such organs as the brain and sensory apparatus, or
the origination, evolution, and diversification of emotions. To
say that these developments *might* have come about through the
selection of chance variation is not evidence that they *did*. In the
case of man, even this purely formal possibility is made dubious by
his inability to tolerate an existence strictly governed by the
canons of biological utility, by his cultural development and
preoccupation with the quality of life in terms of values other than
those of mere survival (of individual, family, tribe, nation, or
species). To think of man as having originated through natural
selection makes it seem that part of his subsequent history is an
unconscious rebellion against the constraints and narrowness
which such an origin implies. The uniqueness of man would
consist in his being *l'animal révolté*, the animal against animality.
But how could the seeds of this revolt have been planted?

[1] Or as though a Martian musicologist visiting the earth in A.D. 2000, and
finding that the only music being made and played was predominantly 'aleatory',
were to conclude that this was how music had always been made.

Temporarily forgetting man as a special and unique 'case', one may ask the Leibnizian question about life: How is it that animals, through their innate structure and inherited styles of behaviour and ways of life, reflect and match their peculiar environment or *Lebenswelt*, from the angle of their special needs and specific ecological slot? Developing animals resemble monads in their early life, and 'dyads' ever after, until death. The genetically regulated or closed programme of embryonic development is the necessary prelude to the animal's responsiveness after birth to external conditions. The eye, for example, develops without specific environmental assistance or outside stimulation, but depends on this for its subsequent functioning. The primary emphasis in early development is on insulated development from within, on 'inner action' rather than interaction with the *Lebenswelt* with which the life activities of animals are subsequently coupled. For many characters, the possibility of a fit between organism and environment is pre-established developmentally (however much the fit may be improved by subsequent interaction and learning). Such characters develop from within, yet match the without; their inner development reflects the conditions and exigencies of life in-the-open. Vision germinates in the dark.

How is it possible that developing organisms should reflect in this way the environment into which they will be thrown, seeing that the latter has no specific influence on the structures that mediate the relationship? These appear to be formed from within, yet informed on the without. Life in-the-world is preceded and prepared by life in-the-womb. A prelude so familiar and in-escapable that its strangeness is lost. It bears a faint resemblance to the theme of an etching by Picasso, in which a cave-dwelling hermit instructs visitors from outside on the ways of the world (quite a different cave-symbol from Plato's!).

The strangeness of these inner-outer relationships is relieved, or rather passed on, through their being ascribed to past evolutionary processes that have adapted ontogenies to the exigencies of open life in a specific environment. But now, on the usual evolutionary story, a similar strangeness recurs, in that the genetic variation on which evolution depends is believed to be independent of environmental conditions, in the crucial sense of not being speci-fically 'formed' or 'informed' by these, so as to 'reflect' them from the angle of the special needs of the evolving group. The gonads are windowless, the genes in their mutations are not environmentally informed or instructed. If the variation matches, this is the sheerest of accidents. Hence the vast importance attached to natural selection. Various palliatives of this accidentalism have been suggested within the general framework of Neo-Darwinism, but none of them seems radical enough or comprehensive enough to change the situation basically. Always the problem recurs: How can changes that are independent match or mirror one another? Can the long arm of coincidence stretch so far as to equip emigrating birds with innate 'maps' of their routes and star-charts? Can it really be thus that their germ-cells have become a kind of coded microcosm of the heavens? ('Seeing the world in a strand of DNA.') Like Hamlet, the shell-bounded chick is potentially king of astronomical space—thanks to its star-informed genes. Astrology itself scarcely suggests stranger connections and correspondence between the heavens and the earth, between stars and birth.

To throw the whole burden of matching on to natural selection is to gloss over the fact that selection can only operate on the basis of variants offered. Selection is the principle of the post-established harmony or fit, but this presupposes a prior supply of the appropriate variation.

Leibniz's question 'How can independent monads reflect or mirror the universe from a particular viewpoint?' may be regarded as a fantastical pseudo-problem. Yet surely this problem, of genetic variation that reflects the changing needs of organisms in a changing world, is the legitimate successor of Leibniz's question. How can independent variables be thus related? How are the cell-imprisoned genes environmentally oriented? Without access to the world, how can they 'instruct' developing organisms on its ways? How have the gametes been matched or 'married' to the world? How is the innate worldliness of lower organisms possible? How has the eye been brought to light?

Leibniz's very way of formulating his question, in terms of the perceptions of monads, seems suggestive for the present problem. This can be characterized, at least in part, as an informational problem. Organisms that are innately equipped for life in a particular *Lebenswelt* may be said to be environmentally informed. Plants are soil-informed, fish are sea- or river-informed, birds may be star-informed, the teeth and digestive systems of animals are food-informed. Organisms are organized bundles or embodiments of worldly information. The specialized organs of information (the senses) make it possible for animals to supplement, during their lifetime, the general information built into their other structures; rather as a mine-sweeper's special mine-sense (sonar) supplements the general 'miney' information built into its design (e.g., into its non-magnetic hull). The organs of information themselves, in their innate structure and mode of functioning, are pre-informed on the nature of the environmental stimulation they will encounter. Thus the perception of the world depends ultimately or originally on genetic variation that arose independently of the world-to-be-perceived. In some cases, the very objects of sensory attention and discrimination, along with the appropriate response, seem to be built into the sensory-motor system—which implies an amazing original coincidence of inner and outer. Once again, selection may 'distribute' such coincidences, it cannot originate them.

The analogy with Leibniz could hardly be closer than it seems here. Where our modern Gonadologies differ from Leibniz—for example, in the strange conceptual mating of causal independence with highly specialized interactions with the environment—this hardly helps to make them more plausible or intelligible. An example of this is that similar effects, of matching the environment, may be brought about by the apparently very different means of a built-in match and a match that is perceptually guided. An animal may be inherently camouflaged, against its normal background, a man may camouflage himself perceptually. In the first case, the perceptions or imperceptions of the animal's predators can only act selectively after the initial resemblance has arisen—as for any subsequent polishing-up of the resemblance. If, however, perceptual information and built-in information are not as disparate as they appear to be, but derive from a common informational 'source', a different account may be possible. The link here is that the possibility of perceptual matching is itself built in, and presupposes environmental information in its evolution. Perception ultimately and originally depends on a non-perceptual matching between evolving percipients and environment. This is, of course, enormously problematical, but it might make us cautious about contrasting too strongly perceptual information and innate information. One might venture to say that they differ mainly in this, that in the second case the environmental information built into non-perceiving and non-learning systems yields no further self-increase during the individual's life, whereas in the case of sensory and learning systems, the environmental information built into them in the course of evolution makes possible an increase in the information accessible to the individual in its wordly career, and ecological transactions. It is the difference between a fixed stock of information, and an ampliative means of getting-informed, in collusion with environmental stimulation. In the latter case the inborn information embodied in the construction of sense organs facilitates the stream of in-borne or empirically acquired information. I do not see how this very subtle and complex interplay between inborn and inborne information could be thought to be accountable in terms of the current doctrine, which insists that all new innate information originates within the germ-cell by random mutation. The transmitter of information need not be its originator. The basic oological argument, that since animals develop from the genetic information embodied in fertilized egg-cells, all this information must have originated within germ-cells, is surely questionable. It is rather like believing that since copies of newspapers are run off printing presses (duplicating errors included), the information they contain originated inside printing offices. The various 'duplicating errors' that have been observed to originate within germ-cells may not be the main source of evolutionary change.

Copying is one thing, originating another. The processes whereby a painting is reproduced, in mass quantity, tell us nothing about its original production. Copying processes are the easiest to mechanize. Insight into the copying processes of heredity, within the species, may be of doubtful relevance for the understanding of the originative processes of evolution. It may be positively misleading. To think of copying and mis-copying as the respective sources of conservation and change, may be an enormous over-simplification, in life as in art! Surely there must be more to the production of new types of life-organization than the faulty reproduction of old types, selectively favoured.

To the question 'What has produced the enormous evolutionary increase of genetic information since the earliest forms of life?' the only honest answer seems to be that no one knows (or if someone knows, he isn't telling). The idea of an endogenous increase of information (by mutation) on this scale, is one that is not, in itself, fully intelligible—quite apart from the 'matching' problem which such an endogenous origin creates, the strange parallelism between inner and outer 'worlds', between intracellular and environmental variation, between micro-cosm and macro-cosm. Surely everyone who has thought about this for a minute has longed to say that the inner changes are respondent to the outer ones, if only some form of interaction was conceivable . . .? It is uncomfortable to follow Leibniz here, without following him all the way, along the sweet but forbidden paths of the pre-established harmony.

This dualism of inner and outer changes (like the dualism of mind and body) makes it seem as if the inner changes were taking place on a different plane—or planet!—from the outer ones. How is it comprehensible that intra-cellular molecular changes should correspond 'code-wise' to the exigencies of life under sun and stars (when these are used as reference points)? What 'nature' has thus sundered from one another, can 'selection' bring together into fruitful relationship? Such explanations are perhaps more amazing than the phenomena.

What if the independence of inner and outer changes, which clouds attempts to understand their relationship, were an essential clue to the mode of genesis or origin of the latter, and even perhaps its necessary condition? This may sound an extreme paradox, but it may be less paradoxical ultimately than to say that inner and outer changes match one another *despite* their mutual indifference. Could not this negative fact of non-inter-action be interpreted as having positive significance, as a vitally necessary bar or 'deterrent' against external interference, or a protective seal?

The causal independence of inner and outer changes constitutes the ideal condition for their matching, on the assumption that there has been a purposive guidance of evolution. By the with-drawal or seclusion, so far as is possible, of these inner changes from the *uninformed* impact of external conditions, scope is left for their informed matching. (Similar considerations apply to the 'inwardness' of embryonic development: the protected embryo is being fitted to interact openly with its post-natal environment when it can do so 'on its own terms', with its reactive systems

beginning to be functional.) This assumed purposive matching of
inner to outer changes would be a way of exploiting their causal
independence so as to bring about useful correspondences which
would not otherwise be effected, but rather hindered, by the
unassisted operation of causal laws. It is, so to speak, the 'in-
efficiency' of efficient causes that provides a reason for postu-
lating purposive intervention. The point may be put in an
archaic-sounding way (but none the worse for that necessarily):
inner and outer changes are separated in the mode of efficient
causation, that they may be related in the mode of final causation.
The separation is vital to the relation. It is its condition, rather
than its 'anti-condition'. The hermit-genes and hermetic nature of
embryonic development are essential elements in the organic
scheme. The embryo is sheltered by its special organic
environment, before being pitched into the open.

For all the great differences between organic development and
man's constructive work, analogies do exist. A boat is made in dry
dock. If prematurely thrown to the waves, it sinks. Man makes
things such as machines, that reflect in their constitution their own
particular machine-worlds or functional environments, from the
angle of man's special needs (or believed needs). He does this in the
absence of any tendency for the unassisted operation of causal
laws to produce these objects. Finality becomes externalized or
environmentalized. This is man's highly original way of 'reflecting
the world'—not *in propria persona* but in his artefacts.

A purposive interpretation of evolution conflicts with the
materialistic or physicalistic philosophy now generally entrenched.
But it is precisely the latter that produces the paradoxes of
Gonadology, of inner changes chancing to reflect outer conditions
and the exigencies of this or that specific *Lebenswelt*. For there are
no known, or even conceivable, physical laws relevant to the
interactions between organism and environment which could
account for the formation of the different organs of the body;
these may be environmentally informed, but cannot have been
environmentally formed through the impact made by external
conditions on the genes, or whatever. Hence, from a materialistic
viewpoint, the paradoxes of Gonadology are insoluble. Hence,
from a materialistic viewpoint, it is necessary to deny that they are
paradoxes, and to fall back on the confidence in time, chance, and
selection that is not fully justified evidentially or analogically, but

explicable as a faith by the general horror of non-materialistic ways of thinking.

The information problem—how environmental information 'gets into' organisms—has two facets already mentioned. The information may be built in, in a fixed and determinate manner, or may be acquired empirically, through the stimulation of sensory and learning systems, which themselves innately reflect certain features of the environment. It seems extraordinary that there should be these divergent ways of being environmentally informed: the way of genetic inheritance, and the way of learning. Such different information-channels! The genes have no windows, yet bequeath window-like information. Why, one may almost wonder, are 'windows' necessary, and organs for sensory learning? The unwindowed information appears to be the more fundamental, since on this depends the possibility of windows, of the innate capacity for perceiving and learning. But this merely underlines the paradox of unwindowed information. 'Unwindowed, but not *unwinnowed*', it may be replied. The information has evolved by natural selection. Even if this be granted as possible, it does not cure the strangeness of the co-existence of these two divergent-yet-convergent sources of environmental information, that complement one another so admirably. However one may think to account for this informational duality, it seems to be of great, and greatly underestimated, significance. I would say it should prompt us to think along the lines of a unitary information source, from which these complementary information systems derive. An unresolved dualism of this kind is surely disquieting. Some attempt to understand their interrelationship is necessary. What makes this so difficult is that their basic principles of information-obtaining (and retaining) appear to be so different. Neither can plausibly be interpreted as a special case or application or extension of the other (though attempts on these lines have been made). Even though the existence of windows on to the world depends on unwindowed genetic information, this does not abolish, but rather draws attention to, the distinction between sensory and non-sensory information; between sight and sightlessness. How can the possibility of sense-information derive from non-sensory information? How can blind evolutionary and developmental processes lead to sight? How can random changes in germ-cells turn ✕

the course of development towards the light and light-sensitive organs? Germ-cells resemble world-scourers and world-scanners and world-mappers, yet without access.... They are prodigiously well informed. One thing at least is clear: sense perception is not the sole source of information on the world. Such a thesis may unconsciously be biased by the fact that its proponents are men, who in some ways are least well informed innately, though best endowed potentially in their learning capacities. If birds and bees were philosophers, they would think empiricism absurd.

That environmental information should be reflected in the molecular structure of DNA, without having been causally 'imprinted' on it by direct environmental impact, is the most nearly Leibnizian discovery of modern science. These far-reaching correspondences between micro-structure and the outer macro-world suggest a unity or connectedness between the organic and the inorganic, the full implications of which might transform our present conceptions of their relationship. If we fully understood how it was possible for a molecule to mirror the world, or some aspects thereof, our understanding of life, its origin and evolution, might be greatly deepened. The ability of organisms and their elements to reflect the world without being causally imprinted by it, might be seen as one of the most significant distinguishing features of living things, whose explanation is one of the deepest problems for any science of life.

Granted that worldly information may be channelled to organisms in different ways, how is one to think of these ways as related? Has each way its own peculiar scope and limitations, which complement one another?

We may take up again the example of mine-sweepers. The information-requirements concerning mines are met in different ways, since there is an independent source of information which informs the mine-sweeper's structures, which manifests itself variably, from the special angle or need for destroying mines without being destroyed by them. Information about the general properties of mines (of various kinds) may be incorporated and reflected in the structures of the vessel; what cannot be built in is specific information about the location of mines or mine charts, as these have no permanent fixed location or predictable motions. Hence arises the need for special mine-detectors. These of course are pre-formed (pre-informed) *vis-à-vis* the general properties of the objects they are intended to detect, but not *vis-à-vis* their specific places. And the places of things are all-important, from the viewpoint of reacting to them. The human fear of insensible dangers is well justified (even though some dangers are increased by their awareness).

Analogously, the environmental information that can in principle be built into organisms and reflected in their structures, independently of learning, is limited by the fact that their reactively relevant environments include mine-like as well as star-like features. Hence arise the possibility and need of both information-channels, which meet in the compromise-solution of built-in systems for the acquiring and retention of information. Thus are organisms pre-equipped to cope with the unpredictable (within limits, obviously). The interplay between innate and acquired information may be seen as a reflection of the relatively stable and enduring features of a specific *Lebenswelt*, and its variability for each individual. The relative rigidity of instinctive behaviour, over many generations, reflects this outer stability, from the perspective of the needs of the species. This unwindowed information constitutes a kind of general species-library, from which all that is of merely ephemeral or individual significance is excluded. For the latter, it is necessary to leave the library or use the windows provided. In the course of evolution, there must have occurred time and again this sorting out of the 'universal' from the 'particular', of the constants and variables implicit in a particular 'way of life'. The pre-recorded tape of instinct is modified, in each individual play-back, by particular interferences from outside. The tape sets the themes, but these are subjected, to a greater or lesser extent, to individual variations. The general themes, but not the particular variations, are handed on from generation to generation; like those traditional chorale melodies on which every obscure organist would once have improvised in his loft, without any chance that his improvisations would be perpetuated.

In the life of organisms, individual improvisation (or adaptability) is necessary owing to the improvidence of hereditary information with respect to the unpredictable details of existence, which cannot be provided for in a general way, in advance, except by the development of learning capacities, pre-adapted to unpredictability, to the contingencies of life, as contrasted with the constancies implicit in a particular way of life. Only thus can provision be made, within limits, for the constant possibility of contingencies that vary from individual to individual.

The 'descent' from the general to the particular, in the transition from inherited to acquired information, is analogous to the descent in human cultural transmission and tradition, when the individual adapts the latter to his own ends. For example, the general rules of language are handed on between the generations, but not the individual's personal and particular uses of the language. A similar underlying principle may be discerned: since no two life-histories are identical, with respect to environmental variables, the hereditary principle is limited in scope, and requires supplementation by some form of individual responsiveness or initiative. Only the animal on the spot can know which way to run. Spot-information, being of purely local significance, cannot be generalized. But the need for spot-information is general, and is provided for by the perceptive faculties common to all normal members of the species. The genetic information on which the possibility of perceptual information depends is information-generating information. From this closed channel there develop the open channels of the senses.

However one may characterize the contrast between these information-channels—as open and closed, windowed and windowless, individual and generic, local and universal, transient and lasting—their interrelationship constitutes a major issue for any evolutionary theory. On top of the question how it is possible for the windowless genes to throw up information that relates to the world beyond, there arises this problem how they are able to complement and correct the inherent limitations of the genetic information-channel by the development of a second channel in direct touch with locally variable situations, capable of providing on-the-spot information. To think the genetic channel capable of 'self-transcendence' in the complementation of its own limits by the provision of additional informational facilities working on quite different principles, is to credit it with hardly credible powers. To regulate the relations between the two channels would seem to fall beyond the scope and competence of either channel. If so, there must be some form of supra-organic regulation and control, which can endow organisms with information via both channels: either as actual information (exemplified by instinctive know-how), or as potential information, made possible by the development of sense-organs and brains capable of learning. For an intelligence, the difference between these two ways of transmitting information, actual or potential, presents no great problem; in our relations with our fellow-men, we easily distinguish between the passing of actual, hard information, and the provision of facilities for the obtaining of information. Sometimes the one, sometimes the other, is appropriate.

All very speculative. One may renounce speculation, if one is willing to stop thinking about evolution. Any theory of evolution is bound to be speculative, in my opinion. (But theories that conform and contribute to prevailing world views are not easily recognized for what they are. Speculation, like treason, 'doth never prosper'.) The basic objection to Neo-Darwinism is not that it is speculative, but that it confers miraculous powers on inappropriate agents. In essence, it is an attempt to supernaturalize nature, to endow unthinking processes with more-than-human powers—including the power of creating thinkers. I find it impossible to share this faith that supra-human achievements can be encompassed by sub-human means and sub-rational mechanisms. Human thought on these matters tends to swing wildly from one extreme to the other. Compromises are unpopular, yet neither the apparent rationality nor the apparent irrationality of evolution seems to me capable of being explained away. This points to a supra-human but not unlimited rationality operative in evolution.

But how is it possible or conceivable that there could be any purposive control of evolution, however limited, now that the physical basis of heredity has been clarified? This is a little like asking how it is possible that human behaviour should be purposive, seeing that the brain is made of atoms. And of course it is not unknown for human purposiveness to be denied or 'redefined', for the reason that it seems physiologically inexplicable. Those who dogmatically deny the possibility of any purposive guidance of evolution are often the very ones who see this as a future possibility, under human control.

It is perfectly true that we are unable to conceive how evolutionary processes might have been guided; but have we a really clear idea how human purposiveness could have arisen or developed from our primitive ancestors, and even ultimately from an inorganic ancestry? To deny the actuality or possibility of any process that we do not understand, or think we understand, would make for great incoherence in our view of the world and of evolution. The real is not necessarily the rationalizable. That some things turn out to be comprehensible may be more surprising than that many things do not.

If for certain purposes it is fruitful to regard Biology as the study of information-systems capable of reproducing themselves in an appropriate environment, then we are ourselves one such system. We are exceptionally well endowed, in some respects, with cognitive capacities, but there is no guarantee that these suffice for understanding their own source or derivation, or the origin of the varying information built into other organisms. On the other hand, it is impossible to know that this is unknowable. The main contention of this chapter is that we haven't got there yet.

The question whether a naturalistic explanation of evolution is possible or probable, is too vague to be useful. Concepts of nature change, and are always relative to human powers of perception, discovery, and conceptualization. What may fall beyond these powers need not necessarily be thought of as belonging to a totally different 'order of reality'. 'The occult' simply means that which is hidden from the senses or understanding of a being with limited cognitive capacities. Occultness is a relative concept. What is occult for one being may be crystal clear for another. Every animal has its well-known (to us) occult zone. It would be absurd to say that what is hidden from an ape, but known to man, falls within a totally different realm of reality from anything that is within the ape's power of apprehension. We may think ourselves into the ape's position, but at a higher level. What is occult for us, may not be universally so. Man may, of course, be the great exception, with no permanent or inherent occult zone. But if he has one, this need not necessarily imply total discontinuity with the world he knows about. The possibility or probability of an occult zone is surely something that a judicious biology of knowledge should draw our attention to. This could not exclude biological knowledge itself.

Is it not strange that the probability of a permanent zone of ignorance or uncertainty should be conceded in Physics, while no such restriction seems to be generally envisaged in the case of the far more complex subject-matter of Biology?

In the case of evolution, an obvious limitation on scientific method is that past evolutionary sequences cannot (yet?) be reproduced at will, under controlled conditions. This makes the testing of an evolutionary theory much less rigorous and decisive than the testing of a non-historical theory may be. This in turn leaves greater scope for the credibility of the theory to be influenced by its conformity or non-conformity to the prevailing mood of the scientific world view. In some ways evolutionary theories seem to me to resemble philosophical 'systems' more closely than would generally be admitted. It is rather surprising that they have largely been left alone by logical positivists in search of new demolition work. Perhaps Neo-Darwinism has been saved from this by its essential contribution to the world view that positivists share.

In conclusion, I would suggest that reliance on some principle of selection, conscious or unconscious, deliberate or automatic, is a fundamental and indispensable stratagem of scientific empiricism, which assumes or hopes that the world is comprehensible uncryptically, without resort to such non-empirical conceptions as purposive direction, or to non-empirical sources of information. The history of scientific empiricism shows repeatedly, as a recurring motif, the invocation of selection-principles to protect the tenets of empiricism, to make the world safe for positivism, and to fend off 'mysticism'. In different domains this defence has been adopted, and doubtless will continue to be. The defence, roughly, is to explain whatever has to be explained as the result of a process of selecting from among given or invented possibilities.

In geometry and mathematics, the applicability, or adaptability, of deductive theorems to reality is attributed to a choice made empirically between symbolic systems. This fends off Kantian or teleological notions, of a mind inherently pre-adapted to the understanding of nature. As in Biology, the saving idea is that of a selection made *a posteriori*, in place of a supposedly rational or *a priori* adaptation to the world. As in Biology, the question how it comes about that the varying possibilities include the ones that fit, is left in a certain suspense. (A kind of 'logical Lamarckism' is sometimes proposed: logical systems which now function analytically, without external check or 'stimulus', may have inherited the results of past empirical learning about the world, which have been incorporated in the premises generating the system. This could hardly explain the applications found for non-Euclidean geometry. So, as in Biology, the principle of selection carried out *a posteriori* between competing 'variations' in logical systems retains its primacy.)

From the above example, a rule of empiricism might be inferred: wherever possible, substitute the notion of selective matching for that of predetermined aptness to the world; or more tersely, substitute selection for anticipation. Such a rule may receive surprisingly varied applications, crossing the frontiers of subjects.

In the field of linguistics, the view that there exists a 'deep' universal grammar that is innate, prior to any linguistic learning (as the condition of learning), is opposed by the empirical conception of the diverse languages of mankind as constituting different selections from the enormous range of possible linguistic variation: selections which are partly free or arbitrary, partly the result of historical development guided by the experienced needs of communication in an evolving physical, social, and cultural milieu, but which in any case owe nothing to any instinctive grammar or innate logical rules. This belief that language is wholly man-made, through man's power of choosing what rules he will follow, in the light of experience, obviates the need for grappling with the difficult conception of a ready-made (or found) grammar, and its possible development through *natural* selection (which would give literal meaning to the old metaphor 'the Grammar of Nature').

Similar issues have been raised in the case of music and the other arts: whether there exist any innate norms or harmonic rules, or whether the artist is free to vary these at will, by an open creative choice. Again, if there were innate norms, the question of their origin and 'selection' would put a strain on evolutionary theory. Men's delight in scenic splendours does not, for the empiricist, signify that somehow they are pre-attuned æsthetically to the appearances of things. Rather, they select or single out for æsthetic approval those phenomena associated with their other needs and interests. Hence the love of harvest scenes, fertile landscapes crossed by streams, and hanging fruit; and the revulsion at scenes of death and decay. Storms and lightning and raging seas are only admired when they do not threaten life, but give one the comfortable thrill of having escaped the dangers they represent. As for the stars that Kant admired so much, they light our nights and guide lost travellers.[1] As men's experiences of things vary, and their active relationships with things, so do their æsthetic reactions. There is no need to postulate some special æsthetic 'sense' or instinct.

In the 'para-normal' field, alleged cases of telepathy, clair-voyance, and precognition are often dismissed by empiricists as specially selected successes, which when viewed against the general run of failures, of deceptive dreams, etc., may be judged coincidental. They thus become comparable, say, with the apparently precognitive instincts of animals, which have been specially selected (in a different way) from the general run of non-adaptive mutations and innate variability of behaviour.

Even normal perception, when 'naively' treated as a process of opening the eyes and registering the scene before them, seems a dangerously pre-adapted and non-empirical form of awareness, and great stress is laid on the highly selective nature of our perceptions, in accordance with our previous learning to dis-criminate, and with our varying needs and interests.

In religious worship, it is a very old idea that the Gods that men adore are their own choice, selective projections of their own features, transcendental mirrorings of man, without cognitive value. If the cows worshipped, their Goddess would be a super-cow.

In the case of moral values and imperatives, any idea of these as coming to man 'from beyond', as prescribed for man, with their own non-empirical form of certitude, may be countered by the claim that they are chosen by man, in the interests of social survival, utility, etc. Relativism, which is closely associated with an empiricist outlook (and correspondingly dubious about the 'inlook' of conscience), may be defined as belief in the ultimacy of varying rules of conduct selected in different social circumstances by local decision or limited criteria that have no general relevance or validity.

It would be too digressive to explore further into the general

---

[1] If they squirted poisonous radiation towards the earth, we would see them differently, in a changed light.

relationship between empiricism and 'eclecticism', or selectivism; or to ask whether empiricism should itself be regarded as a selective, and necessarily partial, philosophy and method of enquiry. In the present context, the relevance of this interrelationship is that any criticism of the use to which selection principles are put, in any case thought crucial for empiricism, can be taken as a criticism of the general underlying philosophy.

In the case of evolutionary theory, scepticism concerning the adequacy of the natural selection principle for the leading role allotted to it can be taken as critical, not merely for the life of a dispensable (and replaceable) empirical hypothesis, but for the very possibility of a comprehensive, empirically grounded science of evolution. Even if this is a false interpretation or inference, adherence to it would help explain the lingering ideological flavour that still persists in some evolutionary discussions. Basic beliefs, disbeliefs, and commitments are involved, extending into many sensitive areas of our culture. Darwinism has stood as a kind of representative paradigm or symbol of the dominant philosophy of our times and our culture.

Does the above analysis imply that the concept of natural selection could be regarded as a regulative or defining principle of scientific empiricism, applied in a particular domain, rather than as an empirically supported generalization? Not quite. More equivocally, it appears to hover somewhere between these alternatives. One could say that it gains an extra measure of support or credibility, of an *a priori* kind, from its apparent indispensability from the viewpoint of scientific empiricism. This is best illustrated when its explanatory power is confidently relied on far beyond the range of its empirical testability, even when such applications of the principle are 'conceptually' difficult or uncertain. For example, when it is used to explain the evolutionary trend towards increasingly complex organization.

2 Language

Voice of a Wood-bird:

'Hei! Siegfried now owns
all the Nibelung's hoard:
if hid in the cavern
the hoard he finds!
Let him but win him the Tarnhelm
'twill serve him for deeds of renown;
but could he discover the Ring,
it would make him the Lord of the World!

'Hei! Siegfried has won him
the helm and the ring!
O! let him not trust
to the falsest of friends!
Let but Siegfried hearken
to Mime's treacherous tongue!
What at heart he means,
that must Mime make known.'

Siegfried:

'Ring and Tarnhelm
when I had seized,
then once again
I gave ear to the warbler:
"O! let him not trust
to the falsest of friends!
To his death he lureth on Siegfried.

> '"Hei! Siegfried hath struck down
> the evil dwarf!
> Now know I for him
> a glorious bride:
> on rocky fastness she sleeps."'

Talking animals make beautiful fables. Nowhere in *The Ring* is there a more charming scene than the above. Could this fable come true? The parrot, notoriously, merely parrots, and from its linguistic incapacity the dictionary has been enlarged. More hopefully, attempts have been made, and are continuing, to teach the use of language to young chimpanzees. If this could be done, the evolutionary gap between human and animal forms of communication would be abridged.[1] The greatness of the gap and the magnitude of the task facing any animal learner, are implicit in the Wood-bird's song.

The Wood-bird, of course, sings German like a native, and answers questions put to it in German by Siegfried. Everyone can hear that during a performance. The business of Siegfried tasting the dragon's blood and understanding bird-song is a picturesque reversal of the 'true' situation. This obvious point is made more obvious by the fact that the Wood-bird's song includes so many features generally held to distinguish human language from the communications of animals.

To notice some of these distinguishing features on the basis of an operatic extravaganza may seem inept, but story-telling is a universal human activity, and language is uniquely fitted for it, by virtue of its 'creativity' which grammarians have recently rediscovered. (A rediscovery may be more important and harder to make than many a discovery.) This creativity implies the possibility of inventing any number of understandable stories never heard before. The ability to follow an unfamiliar story comes close to the heart of the difference between human and animal communications. This certainly seems to sort out the chimps from the boys!

---

[1] Not annihilated. Even if the use of language was transmissible to other species, these would not have originated it. The gap would be much further reduced if animals that were taught language taught it to their progeny; and if they developed the language they were taught along their own independent lines, and invented 'Chimpanzese'.

The same creativity is implicit in the ability to sum up a situation that differs from any previously encountered, and to describe, or prescribe, novel ways of dealing with it. It is apt that a conditional sentence should introduce to Siegfried the idea of becoming Lord of the World, for man's ability to control his environment is connected with his ability to formulate hypotheses and think through the possibilities of change and the instruments necessary thereto. (The Ring is a magic tool.)[1] The use of tenses is essential here, to relate past, present, and future, and to distinguish what has been done from what remains to be done.

Siegfried gets some sound advice from the Wood-bird through the questions he asks. Questioning is an attempt to steer another's speech along certain lines which interest the questioner, though he does not know what the other will say. If he thinks he knows, the question is rhetorical.

Questioning is the way in which unequally informed individuals draw from their fellows the ideas and information that have special interest or urgency. It makes possible the articulation and discussion of problems. There seems no analogy to it in the communication of animals.

Nor is there among animals any very convincing analogy to lying, of which the Wood-bird accuses Mime. If animals could lie, with full intent, they would be moral beings. The notion of responsibility is tied etymologically to the use of language. Infants and animals cannot 'answer' for their behaviour.

No animal, using its normal means of communication, does what Siegfried does, in his soliloquizing before he addresses the Wood-bird. The fact that language lends itself to private or solitary uses, in thinking, remembering, scheming, etc., appears to distinguish it decisively from all known communication systems of animals.

Another important distinguishing feature is the use of metaphor. This is crucial for the creative use and development of language, and its absence from animal communications is associated with their closed, undeveloping character.

---

[1] All utensils have an if-basis: only if one wishes to cook is a saucepan useful. An unused piano is not an instrument but an ornament. 'If ifs and ans were pots and pans': pots and pans exist, not absolutely but conditionally. Change the conditions and their description changes. A pot may 'change into' a child's fancy hat or boot, or an offensive weapon, or part of a machine or work of art. . . .

No animal, in its normal mode of communication, can refer to itself as 'I' or give proper names to individuals. Man's unique self-awareness and group-awareness go along with these features of language.

Man's ability to apply different descriptions to the same individual, which the Wood-bird copies when it describes Mime as the evil dwarf, falsest of friends, and a liar, makes it possible to anticipate and guard against the likely behaviour of other people and objects, in different contingencies. Language makes the future subjectively present, and greatly increases the accessibility of the past.

A striking example of this is man's advance knowledge of death. No animal can give forewarning of death, as the Wood-bird does. It is extraordinary to reflect that so basic a necessity, so manifest a law of life, should be known to men alone, and that they are unable to pass on the news to their fellow-mortals, who obviously have no need of this knowledge in their daily lives. The very term 'mortal', when used as a noun, tends to be used of man alone. The implication of this may be that men alone die in the full sense, with full awareness, whereas animals perish like the grass. Dying as a conscious process and performance is the creation of language.

The thunderclap of this discovery of death, if it had come to the ancestors of man, might conceivably have prevented man's emergence—on the assumption that this news could have proved so demoralizing and oppressive as to weaken the will to live, and led to an early extinction. In the case of men, this possibility may have been averted in the past by the death-adapted reinforcement of religious beliefs, though a contributory factor may have been men's famous ability to mimic the animals and put this knowledge from them, for as long as possible.

If men were able to teach animals the use of language, and pass on to them the news of their imminent death, this would be like a second fall, of animal consciousness, into a death-haunted existence. The consequences of this revelation, for men and for animals, would be unforeseeable.

For better or worse—probably for better—the language gap between animals and men makes it unlikely that the news of their death will spread through the kingdom of animals.

The existence of this gap, while it challenges the understanding of evolution, avoids the imponderable problems of coexistence between animals and men that would otherwise arise. This might be an adaptive gap, a means of distancing men from their close relatives, and of eliminating the possibility of a communion that would be highly anomalous and difficult to regulate, unprecedented in the history of terrestrial life. Experiments in teaching animals the use of language, if they succeeded, would be very dangerous, for all species involved. A basic evolutionary principle would be broken.

From both practical and theoretical viewpoints, success in this venture would be a highly ambiguous achievement. While it would reduce this outstanding gap between animals and men, it would be amazing if animals had developed the latent power of learning the use of language, a power never previously utilized, that could only be released and realized under human tutelage. The creative use of language is an achievement of a higher order than the fixed repertoire of circus tricks, such as riding bicycles and working slot-machines, that animals can be trained to perform. It seems that the only way of reducing the gap that would not be theoretically disconcerting would be to discover a wild group, untrained by man, using conventional signs in a creative way. This, however, would raise further problems: why the cultural developments associated with the creative use of language had not occurred in their case. They would be, like Siegfried when he first acquired the Ring, ignorant of its vast potential. The power of the language ring is confined within the circle of mankind. The attempt to spread this power is probably vain, or if not vain, then dangerous.

Success in this attempt would revolutionize the relationship between species. If animals could be taught the use of language, and transmitted it to their progeny, they would be starting a self-perpetuating linguistic line which would be the product of cultural cross-fertilization between humans and animals. The closest biological parallel to this cultural hybridization would be if animals of different species (or a man and an animal) were capable of producing offspring that were fertile *inter se*. For the animals concerned, this would mean a gross disturbance in their normal way of life and adaptation, that would go far beyond any changes involved in the domestication of animals, into a different 'dimension'. Free cross-talk between species would be adaptively as absurd as free cross-breeding genetically. All existing relations between animals and men, between animals and animals, between animals and environment, would be unpredictably and uncontrollably transformed. The ensuing polyphony would, in all probability, be highly dissonant.

A basic principle of communication, as this has evolved, is that for all their 'internal business', species should have their own private means of communication, sufficiently distinct from those of other species to avoid the risk of interspecific confusion, for example, if the mating calls of one species were liable to attract the members of another species.

This principle of privacy would be breached if different animal species had developed languages sufficiently like man's to raise the possibility of free communication between species. The ability to acquire one such language implies, at least in principle, the ability to acquire another, even from outside the confines of the species. Interspecific parleys, unrestricted in range, would be possible. The liquidity of language—its free flow between linguistically capable groups in contact with one another—would erode existing social barriers between species.

But the sexual barriers would remain. Cross-talk between species, without the possibility of cross-breeding, would result in a weird (and explosive?) mix-up of cultural affinity and association with biological apartness and genetic incompatibility. They would be mental but not bodily mates. Unlike racial differences in mankind, biological differences between species exclude the possibility of de-polarizing innate differences (which go deeper than racial ones) by miscegenation.

The social effect of language is to make possible new modes of cooperation and new forms of conflict. As between different species, the most likely effect would be the enslavement or deliberate extermination of the 'inferior' or weaker species. Free communication between species, in the absence of any real possibility of a common social and sexual life, would be an absurd anomaly and globally disastrous. This has been averted by the species-specific differentiation of communications. Since the signals of a species are potentially public, i.e. capable of being observed or overheard by any animal with the requisite sense equipment, their normal privacy to a species implies that only members of the species can produce them and respond to them appropriately. Man, in his cunning, may imitate the signals of animals for hunting or other purposes, but this is exceptional as an act of purposive mimicry and deliberate deception. Only man can study the signals of other species, and discover their uses. He can break through the privacy of other species' communications as they cannot do for his. This non-reciprocity is a significant aspect of man's uniqueness, of his cognitive superiority.

This exceptional ability to spy on the communications of animals implies that the normal privacy of intraspecific communications depends on their conforming to a general pattern, which human language breaks away from. If the communications of different animal species are normally closed to one another, this is bound up with their being 'closed' in another sense. They are not open to social development and invention for dealing with matters that lie outside the limited range of their inherited functions within the species. They cannot be expanded to take in the topic of other animals' signals—or indeed the topic of their own, these two inabilities being connected. They are tied to certain practical situations. The freedom of speech is absent. So is the freedom to speak about speech.

A relevant uniqueness of language is that only from within a language can there develop that kind of awareness of language, and of other forms of communication, that we call 'having the notion of language and communication'. The social diffusion and development of language depend on a general awareness, however dim, of the nature and uses of language. Near the beginning of this awareness was the meta-word. A 'reflective' language is one capable of self-inspection. Every language has developed some form of meta-language, some way of talking about words and their meaning or rules of use, for teaching and other purposes. Without this sort of awareness of language and non-linguistic communication, man could not have set himself to learn how animals communicate with one another. Lacking this awareness, animals cannot set themselves to learn the communicative signals of other species, and the privacy of intraspecific communications is maintained (man's intrusions apart).

If animals had developed the capacity for the free use and invention of conventional signs, this would in all probability have made possible free-ranging communication between animals of different species, with the consequences suggested above.[1] From this viewpoint, the coexistence of many closed systems of communication, with true language confined to a single species, seems eminently 'reasonable'. It should not necessarily be regarded as a matter of chance. Since it takes two to talk, the existence of a single exception to the rule that intraspecific communications are closed, not open to creative social development, does not have the anomalous effect of a cultural drawing together or convergence of different species that retain their full biological separateness.

This leaves open the bare possibility that on other planets also, a single linguistic species might have come into existence. However improbable or unassessable this may appear, if interplanetary communication at a distance ever became possible, this could not have the same awkward consequences as free communication between different species on the same planet.

Religious people may believe they communicate with God in their prayers and forms of worship and incantation. This raises certain logical problems—could even God understand the arbitrary conventions of a human language without having learned these from the men who made the language?[2] —but these are not relevant here. Everyone agrees that God is a special case.

What these examples may suggest to us is something glossed over when the capacity for linguistic communication is said to be a species-specific trait. Even though this be true, as a matter of fact, it is generally thought to *make sense* to imagine the possibility of free communication between men and other beings, in a way in which it would not make sense to imagine the possibility of free extra-specific communication involving the members of any known animal species, with *their* species-specific modes of communicating *inter se*.

This is a point that might be expanded in various ways. For

[1] This would not be true if different species had developed different 'innate grammars' which made it impossible to learn any language of a different species. Chomsky's main contention seems to be that any language lacking the linguistic universals common to all natural human languages could not be learned so readily by young children. But even on the assumption that an alien language might be impossible to learn, in any circumstances, it is hard to believe in the possibility of co-existence between different linguistic species, with alien cultures. Their mutual incomprehensibility would generate uncanny dread and suspicion.

[2] The mantras or sacred invocations of Vedic hymns, used ritually to evoke the divine presence, were believed to have a supernatural origin corresponding to their supernatural power.

example, if man with the help of language can arrive at general communicable knowledge that is *human* knowledge only in the sense that it is humanly formulated, *not* in the sense that its validity is relative to man's specific mental and sensory make-up: then this knowledge is, at least potentially, a supra-specific achievement, communicable in principle to other beings, without loss of validity. Man would thus transcend the limits of the species. Biological categories could not contain him. His knowledge would be extra-humanly and universally valid.

The difference between the closed communications of animals and the open languages of man appears to be qualitatively such that the latter may more aptly be regarded as deviating, rather than developing, from the former; a uniquely creative deviation, whereby there has devolved on man the historical task of making his own language. (This difference is one that might perhaps in former times have been conceived as the difference between *creatura creata* and *creatura creans*.) The language deviation depends for its uniqueness to a single terrestrial species on the closed nature of animal communications, which leave adaptive space for this single departure from the norm. From this viewpoint the obscurity of the origin of man's language-capacity might be deemed a comprehensible obscurity, since it is hard to think of any way in which the gap could be bridged, or at least of any bridge which did not imperil the principle of privacy for intra-specific communications. This, of course, would still leave completely open the question how this unique deviation came into being. It would be ironical if the language-gap, by virtue of which men can know many things unknowable by animals, should, by its mere existence, draw attention to our cognitive failings, through its inexplicability. Language might then be likened to an invisible source of light.

It must be admitted also that the use of language makes possible many forms of folly that are beyond the scope of animals, who enjoy the lesser fallibility of the uncreative and untheoretical.

It has sometimes been maintained that language, and the cultural development associated with language, make unnecessary further biological evolution. Whether this idea is fully acceptable or not, there is a possible relation between it and the desirability of confining the use of language to a single species. If after they had acquired the use of language the different races of mankind had diverged into separate species, they would presumably have carried their linguistic capacities with them. If the use of language acts as a check on further speciation, this would automatically check the multiplication of language-using species.

It might seem that the devolution of responsibility for the development of language on to local speech communities, with the consequent worldwide differentiation of languages, would encourage in-breeding within these communities, and hence reinforce any racial differences. Language differences would then help to isolate different groups reproductively. Undoubtedly language differences do arouse 'prejudice', do divide men. One need only recall the probable origin of the word 'barbarian' from the Greek attitude to non-Greek speakers, who jabber bar-bar-bar in their uncouth lingoes. Shibboleths may be made out of different pronunciations and accents. But language differences, being the result of learning, are not irreversible, and may quickly be overcome. The children of a linguistically mixed marriage can learn either or both parental languages with equal ease. Language differences are not effective as race-isolating mechanisms. These are socially, not sexually inherited; made, not 'begotten'.

The climacteric view of the advent of language, as bringing to a close one great developmental process and ushering in another, to which no *a priori* term can be assigned, in the new mode of 'creative devolution', gives so far-reaching a significance to language as to make remote any comparison with the signals of animals, which show not the slightest tendency to subvert and transform the evolutionary process through which they come into being. (If they did have this tendency, man might have never appeared. He might, as it were, have been anticipated out of existence.) And even though this climacteric view may be deemed an exaggeration, it is one to which language, and the cultural developments associated with language, uniquely lend themselves. If language be regarded as an invention of man's, no other invention can vie with it in depth and range of consequences, which its originators could not possibly have foreseen, and which even retrospectively are hard to grasp. It stands pre-eminently out, not only from the signals of animals, but from the other intellectual achievements of man (many of which are strongly dependent on language). The very criticism of language, of allegedly inherent 'faults' in it, can only be made through language. If its communicative primacy is ever superseded, this will surely not come about without the help of language.

The languages of the world are encountered as rich gifts from the past, known only in their full potency, equidistant from primal speech, raised up by prodigies of invention from obscure beginnings. A unique form of 'originality', expressed in the original development of language, seems to have been accomplished once for all, unrepeatedly. We may not believe in the noble savage, but we seem obliged to believe in the creative primitive, on whose pioneering accomplishment we unknowingly lean.[1]

Unlike other inventions, it seems to be of the essence of language that once developed, it does not need to be re-invented from scratch—whatever dark ages the race may live through. Without man's extinction, it does not seem conceivable that language should be utterly lost and have to be rediscovered. Man and language are inseparable. It is his birthright, both as a member of the species and as a member of society. There is no known way of stopping children talking, except by physical mutilation or social isolation. Once a language is acquired, not even social isolation will deprive a man of it. It is doubtful if the 'private' uses of language, partly subconscious, admit of total decay and desuetude, even if this were something desired. The indelible experience of childhood, and our intrinsic psychological make-up, see to that. (Brain injuries, of course, are something else.)

At any stage in the history of a language, a necessary condition for its being a language, in common or communicative use within a particular community (which need not, of course, be geographically defined or enclosed), is that there should exist a general consensus on the rules of the language. Considering the complexity and subtlety of these rules, linguistic agreement is far the most impressive example of a general spontaneous consensus developing among men; all the more impressive seemingly, since many of the rules agreed on, especially the semantic ones, appear wholly arbitrary and conventional. 'Conventional'

---

[1] One *might* surmise that primitive man enjoyed a brief period of nobility when he was incapable of lying. Much as one has to learn to follow the rules of a game before one can cheat at it, so the truthful use of words has to be mastered before lying becomes possible. (Alternatively, it *might* be surmised that primal speech was predominantly imperative and hortatory.)

implies a coming-together-in-agreement, i.e., they are rules only because they are agreed to be rules, not because of any intrinsic rightness or aptness. Their *esse est consenti*. Some qualifications are necessary: for example, the extended application of a conventional descriptive term (like 'elbow') to new classes of objects may have intrinsic analogical aptness; but these qualifications need not concern us here. It is social agreement that makes arbitrary correlations the material for meaningful communication.

To say that the consensus arises spontaneously is to say that in general, with rather insignificant exceptions, it is not imposed, organized, or premeditated by some central linguistic legislature. It is not a contrived nor planned consensus. The conventions are not settled by a formal convention, compact, discussion, or debate. (For the conventions of primal speech, this would be unfeasible in principle. The making of compacts presupposes a common language already in use.) The consensus, such as exists at any period in a community's history, is passed on by talking to children or round them, and perhaps by more formal instruction (and similarly for adult new-comers). This is the process of transmitting the consensus, not of originating or changing the accepted rules (except to the extent that that the imperfect learning of children may be a factor in changing the language).

A language is the collective creation of a historically continuous community. In the circumstances, the avoidance of linguistic anarchy is as admirable as it seems surprising. This might well be used as an exemplary exhibit in the anarchist argument, that government is an unnecessary evil and is not needed to make and impose essential social rules, in however complex a society, since language spontaneously reflects its complexities, and contributes to their possibility. The spontaneous language-consensus adds strength to the anarchist paradox (or platitude?) that unruly behaviour is the consequence of the coercive imposition of rules. Anarchism works here, and leads to that ideal combination of social agreement and understanding with individual freedom of expression that has eluded all the governments of the world, at all times and places.

The languages of the world, without central direction, have generally shown themselves capable of adapting to new needs, and have maintained historical continuity without stagnation. The 'internalization' of the rules of language works so well that any sense of their being alien to the individual, forced on him by arbitrary social decree, is normally faint or non-existent. There appear to be universal psychological and sociological aspects of language which reflect a basic human unity or uniformity, persisting through all the other social and cultural, political and psychological differences that divide men.

It is natural to wonder why the growth of language and the processes of agreement and assimilation should be so outstanding and exceptional an achievement, whenever and wherever men are found, and to wonder whether there might not be fruitful political applications of the principles involved. One might wonder also whether these have any ethical relevance—for instance, in suggesting ways of resolving the conflict or alleged conflict between a rule-based ethics and an ethics of 'creative decision'.

Social agreement without coercion—adaptability without planning—universal participation without chaos—individual creativity without loss of comprehensibility—the survival through multitudinous changes of an institution that has no other custodian save the general consensus and consent: in any other field such aims and expectations would seem incredible, the goals of dreamy idealists, whose slogan might be 'Good rules need no rulers'. However—and there has to be a 'however'—belief in the possibility of widening the area of spontaneous social agreement, with language as inspirational model, cannot go unquestioned. For one thing, the conventionality of many rules of language, while it seems to make spontaneous agreement that much the more impressive and improbable, also removes certain obstacles to agreement. The conventionality of language-rules implies that questions concerning intrinsic rightness or fitness, with their capacity for dividing opinion, are irrelevant, and this eliminates at a stroke many possible sources of dissension such as exists in social and political business when there is a substantive conflict of beliefs, sectarian interests, and egotisms. It can never be a matter of life or death, or of comfort or discomfort (except in a vaguely æsthetic way), that one convention rather than another should be adopted. It is impossible to conventionalize all the issues of social and political behaviour, though it may be true that some of these provoke unnecessary dissension, being conventions whose conventionality has, for some reason, escaped notice.

Despite this difference, there remains something peculiarly impressive about linguistic agreement. The mere absence of certain common causes of social dissension does not suffice to explain the universal emergence of a consensus within the varying speech communities of man. This consensus is, of course, never perfect. If it were, languages would be fixed and unchanging. Linguistic changes do not flash through society. Rules are not made or unmade by instant decree. Each of us may have his linguistic idiosyncrasies, with only a slight risk of partial ex-communication. Misunderstandings between speakers and hearers occur for a wide variety of reasons. (Misunderstandings may be fruitful in suggesting new lines of thought, or reducing unwanted ambiguities.)

When all necessary qualifications have been made, it seems an impressive, oft-repeated achievement, that from an apparent infinity of possible languages, each separate speech community has attained and continually re-attains approximate agreement on its own, throughout its historical development. Nowhere else is freedom of choice so vast, or agreement so close, and so little consciously sought. It strikes one as a highly purposive achievement, but also as a very casual one, with no sharp focusing of aims. To borrow (and re-apply) Kant's phrase, a kind of unpurposed purposiveness seems to be operative. Who could hope to trace in detail the social history of a single linguistic invention, in the innumerable transactions through which it passes from mouth to mouth, from mind to mind, broadening or narrowing its range in ever new contexts of communication, gaining new users or losing old ones, as it passes into common currency or fades into disuse? The micro-history of words is perforce ignored by dictionaries. They could never keep up with it. (Moreover, though this is a different point, they are themselves an influential part of that history, not mere recorders of selected aspects of it.)

In the consideration of linguistic agreement, it seems especially notable that agreement should spontaneously be reached on the syntactical organization of a language, without any directing or co-ordinating intelligence capable of imposing a grammatical plan on the users of the language. It is possible that such agreement is not wholly social in origin. If, as Chomsky has strongly argued, there are universal grammatical principles and categories which have an innate basis, these would not need to be socially agreed for any particular language. They would be a necessary constituent of all natural languages. Within a particular language, agreement on these linguistic universals would derive ultimately from innate uniformity rather than social conformity. As between independent languages, agreement on these universals would derive from the innate uniformity of mankind, in this respect, rather than from a common original source (primal grammar), or from the independent recognition that these were the rules best fitted to meet certain universal needs of discourse. This concept of innate grammaticality might be interpreted as the concept of genetic *a priori* instructions, of a formal kind, for the social development of language. It is the concept of language as genetically pre-formed and universally pre-specified, in certain fundamental respects.

While the notion of innate grammaticality may reasonably be considered controversial, it seems likely that some of the opposition to it is motivated by the thought of the strain it imposes on orthodox evolutionary explanation. It would, of course, be naive to suppose that an evolutionary theory could be refuted by its exemplifying certain universal grammatical principles in its formulation! But it would be hard to think of any other evolutionary analogue for this possible state of affairs, in which the form of an invention is innately prescribed, leaving enormous scope for differential invention, such as is exhibited in the different languages of the world. Shall we compare this to a bird-made nest? The form of the nest may be innately prescribed, but not its material. The analogy is too slight to be interesting. Similarly if we say that the form of man-made implements, instruments, and machines is prescribed by innate bodily structures. These do not pre-specify the forms of possible implements in the sense in which innate grammar pre-specifies the form of all natural languages. The analogy that, relevantly or irrelevantly, comes to mind is cultural rather than biological or partly biological: the concept of the pre-invented form of an invention, that invites its own completion and consummation, in an indefinite variety of ways; a concept familiar in music and art. If certain grammatical principles are innate, then by a kind of genetic *avant-gardism*, certain formal instructions and constraints are imposed on participants in the language-game, who are required to improvise a language or discourse on this basis, by a partially controlled

devolution. In *avant-garde* music, the idea is to encourage sponta-
neity and creativity, without total loss of coherence and 'commu-
nicability'—presuming that this depends on some recognizable
pattern and organization. *Avant-garde* concerts, ironically enough,
might be regarded as metaphors or even as ritual celebrations,
for the birth of language.[1] (It is strange how undercelebrated this
great event is, in the arts and religions of mankind.)

But as the formula for a musical improvisation does not, in
itself, constitute a musical discourse, so, if language is innately
pre-formed, this can only come to fruition, for purposes of
communication, through being incorporated and realized in a
particular language, however primitive, with its own socially
agreed rules and conventions. Innate form without particular
content and social consensus would be empty, and the latter
without the former improbable, if all language is founded on
linguistic invariants and innate uniformity. There appears to be
no evolutionary precedent for this, that an inborn capacity should
require for its realization and utilization a particular kind of
social invention encapsulating it, historically developing and
socially transmissible from generation to generation, in parallel
with the genetic transmission of linguistic ability. It is hard to see
how, with any plausibility, this process could be thought to have
been initiated by the selection of random genetic variation, involving as it does the dovetailing of such disparate processes and
parallelisms, and such a relationship as the pre-specified potential
form of an invention to its concrete realization in the essentially
historical task of developing and transmitting a language.

Perhaps only a Lamarck of language could hope to explain this
without the intrusion of non-human rationality or purposiveness.
He could say that the child's facility in language-learning was the
hereditary result of many generations of learning, so far as this
relates to the universal features of language, which might derive
originally from a common source, the rules of the primal language
(assuming that language had a single origin). But how would he
explain the unique creativity of language, and its uniqueness to
man? Somewhere along the line it would be hard to avoid the

---

[1] In some *avant-garde* performances, the metaphor becomes almost explicit,
when names are more or less randomly interjected—e.g. the names for musical
notes. One may also note that a 'music of participation' aspires to the condition
of language, in which everyone actively participates.

assumption that man was originally endowed with a faculty for language, of some kind, and that animals lacking this cannot learn the creative use of language, however protracted their lessons and devoted their teachers.

A general comment on Lamarckism is that, even if it were considered a live option, its potential explanatory power would be limited. It is hardly possible to conceive of a Lamarckian explanation of such major evolutionary trends as increase in brain size. Nothing that the individual organism does with its brain during its active life has any tendency to increase its size, and so the evolutionary increase in size of brain cannot in principle be ascribed to the cumulative inherited effects of brain-use. The whole Lamarckian controversy has limited interest and relevance for some of the major problems of evolution. At most, Lamarckism could only supply a supplementary principle of evolutionary change, within the context of a wider theory. (This corresponds to Darwin's limited use of Lamarckian ideas.)

A very striking aspect of language is its dual transmission to every child (or every 'normal' child, normally reared): on one hand, the genetic transmission of the universal form of language, on the other hand, the social transmission of *a* language, that reflects the local conventions, categorizations, discriminations, and experiences of a particular historical community. Genetic inheritance relates to the invariant elements of language, while social inheritance supplies everything that distinguishes one language from another, by virtue of which separate communities can colonize linguistically and humanize a wide range of varying environments, satisfying their variable needs of communication and developing different civilizations, different cultural continuities.

But this is a duality within what must surely be regarded as a unitary scheme for the transmission of language. Neither makes sense without the other. However disparate as processes, genetic and social transmission have a common convergent goal: the psychogenetic and socio-genetic transference of language and culture from generation to generation. The coming together, in partnership, of these two forms of transmission might be described as a 'covolutionary' intersection or junction—the critical conversion of genetic information and its transmission into a faculty for the transmission of cultural information. (This junction-point is, as it were, the pineal gland of evolution, on an analogy with

Descartes' belief that this was where mental and bodily processes converged interactively.)

This rise of a new mode of transmission and evolution sets man truly apart as its only begetter and bearer. Even though new forms of adaptive behaviour may be socially transmitted between animals, by learning and imitation, this is not truly comparable with the cumulative cultural evolution in which language plays a decisive role;[1] nor can it be regarded as bridging the gap between genetic and cultural evolution, since it is wholly non-symbolic. If the origin of genetic transmission marked the inauguration of organic evolution, the origin of language is the second great inaugural event in the history of life, a fundamental change in the available means of changing old ways of living, in a cumulative fashion. The first of these origins is necessary to the possibility of the second, the second is necessary to the conceivability of the first. The evolution of speaking, and the speaking of evolution, finally as it were converge, in a crowning, if partial, comprehension. And all through the selection of the fitter? Surely there is some deeper significance than this in the way things have gone. Evolution is surely more than a tale told by the fittest.

A. N. Whitehead formulated the concept of different world epochs with different natural laws. Man could be said to have inaugurated a new Earth epoch, in which the forms of the 'old' evolution are transmuted or supplemented. Even if one thinks of the older processes as partly purposive, there remains a gulf between this purposive evolution and the evolution of purposiveness that allows man to participate actively in his own history, with an imperfect but not negligible awareness of his responsibilities as the sole being on earth capable of historical hindsight and limited foresight. I suppose this might be regarded as an evolutionary extension and culmination of the growing sensory awareness of environmental changes that has characterized organic evolution in the animal kingdom; but this awareness is immeasurably deepened by the language-dependent concept of a spatial/temporal 'world' extending beyond the limits of the

---

[1] If Lamarckism were true, organic evolution might be regarded, in part, as the result of cumulative learning and imitation by generations of animals which became genetically 'fixed' in their descendants. Organic evolution would be a series of petrifactions of cultural evolution, and life would be one long learning-game, played against the environment. The players die, but not their winning games.

perceivable environment, and beyond the confines of a 'here-now' consciousness. The change of consciousness is impressively demonstrated by the consciousness of change articulated in evolutionary notions (though obviously it has a much longer history). If organic evolution is thought of as partly purposive, it is not unreasonable to believe that man's unprecedented span of consciousness may be a 'reflection' of this. This may even be its only possible 'derivation'—the alternative being a belief in a mysterious leap into this new dimension or 'logosphere'. It is inadequate to say that this leap is not a leap, but a protracted process of cultural evolution, since one still has to explain the jump from genetic to cultural evolution. If this is attributed to brain-evolution, another kind of leap needs explaining. If it is said that brain-evolution and cultural evolution proceeded concomitantly, interselectively, by easy stages, it is not clear how this would apply in the crucial case of language. At the very least, I do not think one is justified in putting much confidence in these solutions. The problem is an exceedingly difficult one, and there is no answer that is not strongly speculative.

The problem of the transition from genetic to cultural evolution deserves fully as much attention as that of the more famous body-mind relationship, as traditionally formulated. Its depth and importance should make it of equally general interest, not the province of specialists alone. If it was truly a turning-point in the history of life on earth and crucial for the future of mankind, we are still living out its consequences. Compared with it, the events studied by the historians of man are of ephemeral significance, mere details of history, not the ground of the possibility of human history. It sometimes seems as if the more fundamental a transition, the more far-reaching a transformation, the less general attention it receives. Over-specialization makes fools and cowards of us all, since, almost by definition, it inhibits discussion of such transitions. Our modern specialisms, when they reinforce belief in the self-sufficiency of 'subjects' and set up barriers to the study of transitions, are profoundly anti-evolutionary in spirit, though not explicitly.

Even the evolutionary thinking of biologists limits its evolutionary concern when it leaves cultural developments out of account. That genetic evolution should lead up to, or in to, psycho-social evolution, is something to be taken account of in our conception of genetic evolution; as the fact some physiological processes have psychological concomitants is something to be taken account of in our conception of the former. An ultimate and inexplicable dualism between organic evolution and man's cultural history is as unsatisfactory as an ultimate dualism between mind and body. Such a dualism throws doubt on the adequacy of present evolutionary concepts. Coherence fails.

The problem cannot be sidestepped by a refusal to apply the term 'evolution' to man's cultural history. This would not abolish the relationship of the latter to organic evolution. In any case some uses of this ambiguous term seem perfectly applicable to some aspects of man's cultural history. Languages, for example, evolve in the sense that there occur cumulative changes transmitted from generation to generation, which lead to increasing differentiation of the language through time. These changes are, in practice, generally irreversible, though any language may contain 'living fossils' from its distant past, linguistic features that remain substantially unchanged[1] (apart from the 'deeper' universals of grammatical structure, to which the closest biological analogue might be certain invariant features of genetic coding, about which similar questions can be asked, whether they derive from a single point of origin, or arose independently, as the best or unique 'solution').

The problematical relationship between genetic and cultural evolution may be illustrated by reference to the development of a human individual. The genetically controlled prolongation of the childhood of man, relative to other primates, appears to have as its principal function the provision of an extended period for youthful learning, an extension unique to man because useful and necessary only for beings that can acquire a language and be taught many other things, generally with the aid of the language already acquired. This genetically evolved delay in reaching maturity 'presupposes' that an appropriate education will be socially instituted, adapted to the varying cultural conditions of different historical communities, and adapted also to the child's developing abilities, along the lines studied by Piaget. Human ontogeny 'anticipates' pedagogy, and is geared to the acquisition

---

[1] In the word-zoo old and recent forms jostle together without too much friction. Ancient forms of speech may be segregated and preserved in special ritualistic offices.

of language and the cultural heritage partly conveyed by language. *Ab ovo* the human child is culturally predestined or preadapted, though only in the most general sense. The 'influence' of language on human biology goes deeper than the provision of special language-learning capacities (and probably in other, more negative ways, such as the almost complete elimination of innately fixed patterns of behaviour).

It is scarcely necessary to mention the profound psychological and emotional implications of a long speculative adolescence, or the significance of this for man's creative impulses and idealistic aspirations and desperations. The organic processes of development have come to be invested with supraorganic, cultural, and evaluative significance, in an unprecedented way; a transfiguration of animal growth, not without its disproportionately emphasized drawbacks and disturbances, but by no means confined to these. While it is an evolutionary commonplace (if anything about evolution is commonplace!) that organs originally developed for one set of functions may acquire new ones in their subsequent evolution, these new developments in the field of individual ontogeny appear to be in a class of their own. The phenomenon of play is, admittedly, common to men and animals, but only man can convert childish play into art (and back again sometimes) or experience a cathartic involvement in a tragic mime (or a Brechtian detachment therefrom).

One may surmise that the cultural value of a prolonged childhood (to be balanced against the disadvantage of protracted dependence) has favoured a kind of Peter Pan selection, in which those who grew up too quickly failed to acquire their full cultural heritage (or a substantial part of it) and were disadvantaged thereby. But such selection could not be expected to operate without the development of more or less specialized learning capacities in the young, or in the absence of a substantial cultural heritage to be acquired. It seems doubtful whether these apparently coordinated developments can be adequately explained, in principle, as a result of the selection of random and uncoordinated genetic changes. The complexity of the cultural transformation of man makes such explanations seem rather facile.

Man's cultural evolution, while dependent on the older kind of evolution, goes beyond that and invests it with new significance, as the individual's development, while dependent on the older processes of ontogenesis, deviates from these in its supraorganic significance. Individual development has been slowed up, though social and cultural evolution may go at a far quicker tempo than the older kind of evolutionary change. Organic evolution, as currently conceived, leaves room for this neo-evolution, as organic ontogenesis leaves scope for this new development in 'growing wiser'; but in neither case is the transformation effected in the older forms of development fully comprehensible. There is a hiatus here that we simply do not know how to account for. The processes of organic evolution, in making possible this neo-evolution, appear to have surpassed and transcended themselves in a truly remarkable way.[1]

This surely is the central issue on which evolutionary discussion should fasten, in an undogmatic way, as befits this most open of questions, on a matter of fundamental and universal significance, involving as it does the whole relationship of man as a historical being to other beings in their temporal relations, both onto- and phylo-genetic, and their ways of living in a changing world. When opinions on this issue become fixed, they become most questionable, through leading the protected life of the unquestioned. No theory should be turned into an institution, centenially celebrated. When a theory anæsthetizes our sense of surprise, or even of shock, that plays so necessary a part in sensitizing the dormant mind, it removes the stimulus necessary for its own application, appraisal, and development. This is not the fault of the theory, but of its reception. Always to be resisted is the tyranny of 'today'—'we are all ——ists today.'

The fact that this cultural breakthrough has isolated man, whereas the older kind of evolution relates all forms of life in a single web, enhances the freakish ambivalence of man, as an animal among animals, subject to the same basic processes of development, reproduction, and assimilation, and one who stands alone, in his capacity for independent self-development in ways all his own. There is no 'animal function', or at least none that man becomes fully aware of, to which cultural or symbolic significance may not be given. Even in his sleep, man dreams

[1] Though one may feel a sneaking sympathy for the science of Pataphysics, whose founder, Dr. Faustroll, maintained that the world consists of nothing but exceptions, the point is that man appears exceptionally exceptional, in finding and following this new evolutionary path.

culturally, and the most extreme derangements of the night may be fed back into his life and his art; a new 'function' indeed, for this ancient organic respite! Night-world and day-world interpenetrate, in novel ways. The most fleeting dream-wisp may be converted into a durable monument. Man refuses to be limited by the organically evolved uses of his 'natural functions'. He transposes these at will, using foot as hand or vice versa, and in other ways subverts the organic order as this has never been subverted before, for example in the practice of bodily deformations, culturally prescribed. He is unwilling to accept that his body, as it comes to him from God or Nature, is unimprovable. His attempts at restyling set him up in direct competition with the processes of organic evolution. *Nihil tangit quod non subornat.*

One feature of the relation between organic and cultural evolution is the ambivalence, or 'ambivaluence', of the latter with respect to its typical productions and inventions. Some of these can be viewed as serving similar adaptive functions to those that have been promoted in the course of organic evolution, by a kind of devolution and displacement of adaptive energy, which now is primarily directed at external adaptations, the so-called 'exosomatic organs'. But there are other aspects of man's many-valued cultures which cannot reasonably be regarded in this light: they have no evident biological or social utility. The new forms of adaptation facilitated by social and cultural organization make it possible for man to follow his insight that the mere continuation of existence is not, for him, 'enough', or good enough. Again we encounter the very perplexing mixture of continuity and discontinuity with the animal kingdom. There is the primary discontinuity of man's cultural development, and within this, there is both continuity and discontinuity with respect to the ends that are culturally promoted and transmitted.

Language itself is a most useful device, which comes to be valued for far more than its social convenience, or its indispensable contribution to practical reason and problem-setting. In many ways language has created new imaginative possibilities that cannot be subsumed under any general formula. Even if all our biological wants were instantly and miraculously satisfied, we would still find in language an inexhaustible source of interest and exploration. The prisoner thrown into solitary confinement takes with him the accumulated luggage of language, the contents of a life enriched by language, which can be summoned up through language, together with the play of mind that language assists.[1] If we were promised eternal life after death, but without the power of language, we would doubt how 'meaningful' such a life would be. We would feel like poets deprived of words.

A Darwinian could easily maintain that the wide employment and enjoyment of language is in no way incompatible with the belief that the 'language faculty', whatever its true nature, evolved through natural selection.

This belief cannot, of course, be conclusively refuted. It remains possible that man's innate flair for language takes a less specific form than the prescription of specific grammatical principles and categories believed to be common to all natural languages. Innate grammaticality might be a more general, less closely defined aptitude (resembling innate musicality, possibly). The existence of linguistic universals, even if well established, does not necessarily entail that they are genetically inherited. As for the 'infant prodigy' argument, that children are marvellously adept at learning the grammar of any language to which they are exposed, there exist infant prodigies in other domains, and the same argument ought to apply to these; but in our present state of ignorance, it would be rash to affirm positively that all infant prodigies were innately possesed of the principles and basic 'grammar' of the activity at which they were prodigious. Between the two extreme theories of knowledge, represented respectively by Plato and Locke, lies a range of intermediate possibilities. Modern linguistics does seem to support a movement towards Plato (or Kant), away from Locke. This has to be taken account of, in any evolutionary theory concerning man and his language. The prodigality of infantile linguistic talent should be the concern of such theories. Until the nature of this talent is clarified, the question how it might have developed cannot be adequately formulated.

Even on a more general interpretation of man's language faculty than that offered by Chomsky, some of the difficulties already mentioned that seem inherent in the Neo-Darwinian analysis, would still persist. Further, one may well ask how, in this case, the selective process could be supposed to operate. If the language

---

[1] Any writer sentences himself to long spells of solitary confinement. From his solitude a new 'world' may grow.

faculty evolved on the basis of random genetic mutation, it would have no selective value until it had been socially expressed and developed through linguistic invention and convention—the necessary agreement on the rules of the primal language, however simple. The mutant genes would not increase in relative frequency through selection, until this language was in use, and improving its users' chances of survival. In other words within the original population there would emerge a sub-group of language-users, an 'inner circle', that did not owe its initial linguistic ability and achievement to natural selection. If a ring of language-users can arise independently of selection, the reasons for thinking that selection is necessary for enlarging the circle and spreading their 'talents' appear to be weakened. If a sub-group within a population can be pre-adapted to the use of language, without benefit of selection, why should not a similar pre-adaptation be conceivable for the population as a whole?

The ideal assumption for Neo-Darwinism would be that the whole notion of an inborn language faculty is a fiction. If the use of language could be regarded as one fruit of a more general mental ability, one might assume that this ability had been selected, in partial independence of its special linguistic application. But when one tries to think of other applications, such as the invention and sculpting of tools, it seems fairly evident that the creative use of language calls for special talents not needed for other inventions that might be thought to antedate the use of language, or to have arisen concurrently with the latter.

3 Beyond Language

Every human faculty, like every human art or science, gives rise when critically reflected on to an awareness of its own limitations, an awareness inseparably connected with the exercise of that faculty or art or science and growing out of it. The power of language makes certain kinds of questioning possible, which reflexively leads to the questioning of the power of language and an awareness of possible limits. One such limit is summed up in the concept of indescribability, which could not be arrived at except through language. Language so to speak un-languages itself, points beyond itself to a region inaccessible to itself. If anyone were to deny that anything could be known to be indescribable, on the ground that we could not know of anything indescribable except by describing it, this would resemble the denial that invisible entities could be known to exist, on the ground that perception gave the only proof of existence. There are few positivists left today so positivistic as to deny that there can exist a science of invisibles. On the other hand, it seems obvious that there can be no adequate science of the indescribable and immeasurable, as scientific description and theorizing depend on linguistic classification or mathematical specification. The limits of language and measurement constitute the limits of possible scientific knowledge.

Can we know of such limits to knowledge? Can we know that there are indescribables? I think we can, at least if we take 'indescribable' in the slightly weakened sense, to refer to something of which all possible descriptions can be recognized as seriously inadequate and defective. Leaving aside other possible indescribables and silencers, I think the least contentious case might be that of musical compositions, and the experiences associated with their making, playing, and hearing. If music were capable of adequate verbalization, it would lose its uniqueness and indispensability as a necessary form of human expression and communication, whose loss would be irremediable and incapable of mitigation by any verbal (or other) substitution. Indescribability, or untranslatability, is the necessary condition of the unique value of music and the other non-linguistic arts, and this limitation of language may therefore be welcomed, since without it human awareness would be greatly monotonized and impoverished. The wish to find a verbal equivalent to a piece of music, to make it 'intelligible' verbally, is a music-killing wish. This is the aggressive (or 'reductionist') aspect of the apparently innocuous, if grandiose, aim of finding a theoretical (verbal or mathematical) explanation of all human activities and forms of experience. It seems that this could only be achieved by distorting or denaturing experiences that are language-resistant, and expand our awareness and forms of communication in ways that language can neither follow nor emulate. The limits of language are the life of music. Long live music, long live language—and never the twain shall fuse.

The verbal-mathematical perspective of science and philosophy makes it easy for their exponents to forget that there are non-verbalizable, non-computable forms of experience and expression. This forgetfulness is one of the factors leading to 'scientism'. The case against universal verbalizability can only be apprehended non-verbally. It is impossible to explain verbally what it is about a Beethoven quartet that defies verbal rendering. There are more things in heaven and earth than are dreamed of by your philologist, which are necessarily missing from his universe of discourse, and from the Leibnizian dream of a universal, all-comprehending language. Life is too rich for that. It is the overlooking of this wordless richness that is the 'reductive' error of the exclusive verbalist or rationalist. The 'demand' that everything be verbalizable is as unreasonable as the demand that everything be paintable or sculptable or musicalizable. Mahler's aim was to pour the whole world into his symphonies, but his musical universe is not the one described by the literary or mathematical cosmologist.

The verbalistic bias common to philosophers and scientists, at least when they are engaged on their own work, shows itself in the belief that thinking and the use of words, or mathematical symbols, are closely conjoined if not identical. Yet no musician doubts that there is a musical thinking that does not rely on words or issue in them, that is concerned with the organization and development of coherent musical 'ideas', and with the solution of musical problems by purely musical means. In view of this common usage, it seems arbitrary to say that musical thinking isn't really thinking! Those who write about thinking and relate it closely to the use of words are, as it were, preset in a verbalistic frame of reference. Probably most writers are predominantly 'verbalizers'.[1] Verbalization has the same kind of dominance in our thought about thought, as vision has in our thought about imagery and perception. Though this dominance may be partly justified, at least from the viewpoint of our practical and theoretical engagement with the outer world, it should not become exclusive and totally domineering. There exist a tone-thinking and a form-thinking and a colour-thinking, as well as a verbal thinking. However difficult this is to understand verbally, there is overwhelming testimony that some music may properly be called meditative and reflective, and a fair amount of agreement about what pieces fall into this category. There is a recognizable musical pensiveness and musing, which assures us that music is after all an issue of the same mental apparatus that in other media produces different forms of constructive thinking. Between such media or forms there are both diversity ('untranslatability') and underlying connections and unities, which can be experienced and exhibited but which are not perhaps clearly and articulately statable.

That musical thinking is so to speak self-fertilizing, rather than mated to some particular alien object or situation or dilemma, does not destroy its thinking-character, any more than is the case with mathematics, similarly mated to itself. Here lies part of the affinity between music and mathematics, though presumably there will never be constructed a *Principia Musica-Mathematica*! Still, analogies can be discerned. That some music may be turned to 'programmatic' purposes is analogous to the applicability of some parts of mathematics. The æsthetic distaste for programme-music, and the preference for 'absolute' music, may be compared with the strictures of a G. H. Hardy against applied mathematics and his desire to keep mathematics pure. This preference for objectlessness may be compared again with the preference

---

[1] Think of the title of Sartre's account of his early life, *Les mots*.

for non-objective painting that can follow its own æsthetic adventures, unimpeded by object-slavery. (Compare the phrase 'slavish imitation'.) It is not necessary to underline the connections between these valuations, connections that show the working of a unitary mind in different fields and media. But these fields retain their differences. Music will never be 'reduced' to mathematics, nor mathematics to music. To call a fugue 'audible mathematics' merely underlines this. There is no theorem in the Theory of Numbers which requires a keen ear for its full appreciation (or a practised 'inner ear', to match the case of the expert score-reader).

The adventures of the mind into verbally inaccessible regions should keep us perpetually mindful of the limitations of language and science, and warn us that scientific empiricism, when taken as a whole philosophy of life, is a false impoverishment of the full range of human thinking and experience. In short, its empiricism is defectively selective, however unavoidable this selectiveness from the verbal-mathematical viewpoint. A 'narrow' intellectualism is narrow by virtue of its indifference to non-verbal and non-verbalizable processes of thought and communication. Scientific reductionism can be explained in part as the result of a totalizing fallacy, of equating the sum of significant thinking with all that is or could be thrown into propositions of a verbal-mathematical kind. This is a very 'bookish', perhaps even Mallarméan fallacy—that a book could summarize the full range of human thinking and experience. Mere literacy and numeracy are inadequate as totalizers. It may of course be true that only a being with the mental equipment that lets him be verbally communicative and creative, can also be non-verbally or hyper-verbally creative. There are profound analogies, and profound disanalogies, between verbal and non-verbal forms of thought. In this situation, it is wholly intelligible that a metaphor such as 'the language of art' or 'the language of music' should both attract and repel. The exploration of this metaphor is still in an embryonic stage, and should receive far greater attention from linguistically minded philosophers.

If this metaphor be used, an instant qualification must be that music is a language without names conventionally attached to their objects or referents. This helps us to understand why an accurate verbal translation of music is not possible. This verbal impotence is not due to some contingent and rectifiable failure to find or invent the right words that would make music fully accessible to the unmusical, the tone-deaf; rather it is inherent in the nature of music that verbal translation is not possible. Translation from one language to another is made possible by the conventional nature of the relationship between words and what they are about. When conventions are substitutable, translation is possible. Translatability is an unintended consequence of the conventional character of verbal signs. This engenders the possibility of an indefinite number of possible languages, which engenders the need for translation, a need that is partly satisfiable, at least, through that same conventionality which creates the need, by its multiplication of languages. If all verbal signs derived their meaning from resemblance between sign and signified, there would be little or no need for translation, since all languages would resemble one another, at least to the extent that they all reflected similar 'life worlds', with similar thematic interests in communication. Convention-based languages have far greater communicative scope and freedom than purely imitative language would have, and this more than outweighs the disadvantage of conventional language, in the dangers of misunderstanding and the extra burden of learning that it imposes.

The absence of conventional definitions and semantic rules in music does not preclude other kinds of conventionality, and these may present certain analogues to verbal translation. When a piece of music can be transcribed for different instruments without its 'musical substance' being greatly affected, this implies that there was a large element of convention and arbitrariness in the original choice of instruments. The smaller the change in musical content, through transcription, the greater the conventional element in the original scoring.

Translation depends on the availability of conventional synonyms. This term is often confined to words of the same language, but translation between languages may be regarded as an extension of paraphrasing within a language, rather as the state's conduct of external affairs may often be seen as the extension or translation of domestic issues and policies.

It is possible, and even customary, to speak of different musical languages, but these are not related in the way that Greek and Latin, for example, are related: they are not intertranslatable.

Elements of different musical languages may be hybridized, for example particular sonorities and harmonies and forms of musical syntax, and this is analogous to the way in which different verbal languages borrow from one another. But such borrowing does not greatly strengthen the specific analogy between music and verbal language, since it is common to all forms of art and communication, and indeed to all forms of activity in which cultural borrowing or stealing goes on.

Translation, of course, presents many problems even for convention-based languages. Apart from the paucity of close conventional synonyms, it may be difficult or impossible to reproduce nuances of style and tone, or 'atmosphere', or charm and toughness, or verbal conceits, paradoxes, and puns. Poetry is considered pre-eminently untranslatable, and may be compared with music, especially in its use of metre, rhythm, accent, and assonance. These cannot be translated, in the usual sense of giving the same meaning in different words, though attempts may be made to reproduce them, by imitation, in a different language. Poetry still differs from music in being made out of words, and this in principle offers the hope of the possibility of translation, however 'loose'.

Since music cannot be verbally translated or paraphrased, or given an equivalent 'meaning' in some other medium, musical 'understanding' is *sui generis*. How musical communication is possible, in the absence of anything corresponding to the agreed definitions and semantic rules of verbal language, and the paucity of musical 'imitations', is deeply mysterious. But a vital clue to the ravishing power of music and its potent, largely ineffable influence over states of mind, may lie in its paucity of conventions of the above kind, arbitrary and socially agreed. Music somehow gets at us in a direct, unmediated way, with a minimum of convention, of conventional substitutions, tokens, and inter-mediaries. We may possibly be pre-attuned innately to respond with moved attention, thought, and feeling, to particular sound-structures that have no conventional significance assigned to them or their elements by definition, provided these form recognizably part of a musical style and syntax that we are familiar with, and provided we are not afflicted with tonal deafness, or some other impediment. There are significant differences of musical culture, and between cultures. But musical communication does seem ultimately to depend on mental similarities between 'senders' and 'receivers', pre-dispositions for responding to the power and glory of music. These alone can facilitate a musical response for the solitary (maybe deaf!) creator of tonal systems that have no conventional anchorage in a common public world. Our musical perceptions have the remarkable property of abstracting us from this common world, in opposition to the normal function of the senses. Music is the exploitation of the auditory sense for essentially 'unworldly' purposes, an evolutionary change or addition of function that takes some explaining! Some innate attraction and affinity has developed making us revel in sound and its richness of expressive import. Sounds become the docile servants of communication, but only when ordered and ordered about in special ways for which the formula does not exist. Hearing has become the most human or inter-human sense, for purposes of communication. No man can be so humanly isolated as the deaf one. Hearing is *par excellence* the sense through which states of mind are conveyable, so that 'the soul becomes audible'. This can give intense satisfaction and joy.

Music can of course be used to mimic the sounds of nature, but this is its most dispensable and least characteristic power, though sometimes it can come as welcome relief from too intense a meditation. But the limitations of music in this respect reveal by contrast its extraordinary subtlety and fertility in its non-imitative uses. No one can give an adequate verbal analysis of what is meant by musical profundity, but a musically perceptive person usually knows it when he meets it, or at least after several meetings. One comes back to the inadequacy of words in the discussion of music, and the inherent weakness of musical criticism, compared with literary criticism. One of the best services that words can render to music, is to clarify their own ineptness in this field, and thus let music appear in its true light, as a unique and verbally inexplicable form of communication and experience. As was suggested earlier, this lack of verbal understanding, if it is resented as an insuperable limit, is a small price to pay for the unique richness of musical experience. The demand for verbal comprehension is a fundamentally unmusical demand. What would one think of the reverse case, of the demand that all verbal sentences should be musically translatable or intelligible?

Since words and music are the issue of the same mind, in its attempts to communicate its thoughts and experiences, they are not wholly disparate, and can be yoked in mutual pointing; but this yoking is very different from the translation of sentences from one language to another.

This underlying human unity in different forms of expression does not preclude an irreducible diversity. Man's mind and creativeness are too rich to find full expression in any one medium. Too verbalistic an education is miseducation concerning the powers of the mind and the sources of ecstasy. Other biases would be similarly censurable. There is no genius and no art so universal that the full range and potential of the human mind is manifested in him or it. The genetic replication of genius, if it became feasible, would be doubtfully desirable. Who wants a Beethoven or Wagner duplicated?—or multiplied!?[1] *Identica non multiplicanda præter necessitatem.* The duplication of an unfulfilled genius, dying very young, might be desired; but the result of such a doubling would not be foreseeable.

The prevalence of the human passion for music shows as clearly as anything that man through his innate affinities enjoys some independence of his external environment and that his mental make-up has not been wholly determined by adaptive needs and constraints. If this assurance of man's partial independence of his physical environment be one of the functions of art, the claim that music is highest of the arts could be justifiable. No other art enjoys the same degree of independence, however envious and 'aspiring' it may be towards music. The virtual elimination of the physical world as part of the subject-matter of music gives to music its unique creative freedom; and also possibly its ecstatic quality, through this possibility of 'standing out of' one's normal, 'worldly' self. Total absorption in music may be compared with absorption in mathematics, but the former covers a far wider range of human experience, and does not have to prove anything.[2]

Thus it is precisely in one of his greatest, and least environment-bounded powers, that man is not fully comprehensible from a scientific, verbal, or mathematical viewpoint. The scientific value of verbalizability is checked by the alternative human pursuit of

---

[1] It does not yet seem to have happened by chance that a person of outstanding creativeness has had a genetically identical sibling. It is impossible to predict the outcome of such a double incarnation. (What if Christ had had an identical twin?)

[2] Except 'on the pulses'.

non-verbalizable values and experiences. These contribute largely to a sense of the value and meaningfulness of life. This helps to explain why scientific accounts of life leave a sense of its total meaninglessness.

So long as there remain non-verbalizable and non-measurable thoughts and experiences, no scientific theory of man and no philosophical world view can be other than partial and fragmentary. What these necessarily leave out may be as 'significant' (whatever the criterion of this) as what they include. There is absolutely no reason to believe that language can take the full measure of things. This is an insight we can only reach through language, which alone can question its own omnicompetence. The greatness of language is shown not least in this. Through its faculties for self-questioning and self-limiting, it guards us against the idolization of language. Through language one can acknowledge that man as the author and recipient of non-verbalizable experiences can go beyond, or 'outside', language. This is an empirically verifiable principle of verbal indeterminacy. A consequence of this is that such experiences are in principle unpredictable, except in the loosest possible way. This adds greatly to the interest and richness of life. A Laplacean life would be terribly soporific.

There is no more impressive witness to the greatness of the human mind, and whatever brought this into being, than that it should have developed these two complementary means of communication and contemplation, language and music, the one best adapted to discoursing about the everyday world and its imagined extensions, the other to the communication of inner experience, with a directness and sometimes shattering intensity all its own. In our language, as in our music, we are constantly making surprises for other people, and for ourselves often. It seems as if there is an inexhaustible amount to be communicated, in man's prodigal universe of discourse. An art may go into decline for long periods, but will revive in some other century or locality. For a people to be wholly artless is as unlikely as for it to be speechless. In neither case is it a question of an optional ornament of life.

A man has the outward appearance of an animal but the brain of an artist. This is one aspect of the human paradox. (Of the human disguise, one might almost say.)

We do not know what it is that distinguishes the brain of an artist from the brain of a non-artist, and we do not know how it came about that the brain of an artist was developed from the brains of non-artists, in the surprising course of evolution. We do not know, and there is no guarantee that 'we', or our descendants, will ever know. Furthermore, if æsthetic experience is incapable of adequate verbal or mathematical specification, can it be safely assumed that the cerebral processes associated with such experience are in principle specifiable? Might not the principle of verbal inadequacy or indeterminacy have a wider application? It is certainly hard to see how there could be made any precise statement of the cerebral-æsthetic association. This must be left here as an open question for the future. We should possibly expect to find unexpected barriers to knowledge, especially in this field. Subjective and objective descriptions are so intertwined that any major gap in the one is liable to affect the other. Gluck may be at least as relevant as Gödel to the question whether the human brain is in principle capable of full self-understanding. The injunction 'Know thyself' may well be unfulfillable, for a variety of possible reasons and interpretations.

The question of the origin of man's artistry and passion for music seems as undecidable as any, in the backward abysm of time. If one thinks of music as a language without names or objective referents, one might suppose that it derives ultimately from a primitive preverbal stage of communication, and has retained the directness and primal intensity of this. But such speculation seems futile. Music is a marvel of whose origin we are destined to remain ignorant, and whose innate source (in some pre-established love of harmony?) is the obscure glory of our race. Love—this is is the word missing from our so-objective accounts of human evolution. True to himself, man comes to love most that which distinguishes him from other beings: not in self-love, nor in *ars amoris*, but in *amor artis*. This is a love that passes all understanding in narrowly functional terms. It is a source of sheer surplus value, though not without its trials and exactions.

Man here, as elsewhere, outstrips his self-explanations. *Homo homini lepus*—man out-hares man, being in advance of his theories about himself. We should never take these partial accounts for more than they are worth. Man's unique creative powers, especially in fields that are language-resistant, limit his powers of explanation of himself. Hence any theory that claims comprehensiveness will be reductionist, or simply inadequate. This is not obscurantism. Limited explanations are fine, when their limits are openly acknowledged.

Men's creative energies in the fields of art and music expose the limits of the adaptive explanation of human evolution and behaviour. If one said these were as necessary to human life as food and drink, this would imply a supra-biological concept of human life and its needs. It seems futile to seek to deny this, in order to protect and preserve a pet theory, for instance by suggesting that art may be a side effect of man's unique and very useful capacity for foresight and prediction. A strange by-product that turns against its source and makes foresight impossible. In some inexplicable way, man in his art seems to be following partly in the steps, not of his animal forbears, but of the evolutionary process itself, by the making of unforeseeable transitions and unpredictable novelties which can only be spoken of after they have arrived, and which make surprise one of the most pervasive and appropriate reactions to the human and non-human universe. The capacity for surprise goes along with the capacity for foresight, and is inseparable from it.

Creativity and predictability are antithetical conceptions, if one defines 'creativity' as the power of producing novelties by no known rule or formula, the application of which might in principle have predictable results. Hence the search for predictive explanations and formulas may be anti-creative, in effect if not intent. We may here be up against incompatible human goals and aspirations. These can in principle be reconciled, provided limits are recognized in the pursuit of the clashing values of creativeness and predictability.

The main contention of this chapter is that scientific theories about man are often unconsciously selective and biased in their choice of material, owing to the requirement that the latter should be clearly and adequately specifiable in words or mathematical symbols. Theorizing has its necessary conditions that are legislative for the possible subject-matter of theories. Hence all scientific accounts of human existence are partial, dealing with pre-selected aspects only. 'Scientific man' is a linguistic artefact or abstraction. To confuse him with the fully human is to be beguiled by language and its limitations. These limitations *seem* to be inherent, and irremediable.

There is a useful ambiguity in the phrase 'scientific man'. He is man made in the image of the scientist (another abstraction!) to the extent that he is indifferent to the non-verbalizable and non-rationalizable aspects of experience.

Such single-mindedness (comparable with Blake's 'single vision') has its achievements, but a sane philosophy would set these in the context of the full range of human creativeness and experience, whether rationalizable or not.

4 Consciousness

How is consciousness possible? How is it that one's children are conscious, and one's car is not? So far as is known, there is only one way today in which a new conscious being can be 'made': namely through being sexually propagated by other conscious beings. Like the parallel principle, *omne vivum ex vivo*, this one leads back to the very difficult question of remote evolutionary origins.

So far as concerns the rebirth of individual consciousness, it seems reasonable to presume that this is genetically transmitted from parents to children; or if one prefers another way of putting it, that infants have certain sensations and feelings of discomfort through an inherited predisposition to experience these when their bodies are affected in certain ways. In some cases, the genetics of consciousness is well established, for instance, in the inheritance of different forms of colour-blindness, which entail restrictions on normal sensory awareness and discrimination. Consciousness, in this sense, Mendelizes. One sees the world, not through one's parents' eyes, but through the eyes, etc., made possible by a particular mixture of parental genes. This may prevent the development of whole modalities of awareness, but total and permanent anæsthesia is incompatible with human survival, except at a purely 'vegetative' level.

Without this constant regeneration and reproduction of consciousness, the principle of vegetation or automation would reign supreme, in the all-indifference of all to all: a state of affairs widely believed to have obtained in the remote past, and liable to recur, possibly; though with a rather frequent, if illogical, asymmetry, noticeable in other domains also, the prospect of total insentience may appear more disturbing than its retrospective consideration.

The regeneration of individual consciousness (human or

animal) during development is an occurrence that is comparatively neglected in the enormous library of writings on other aspects of the mind-body problem. As has been stressed by C. H. Waddington, hereditary and evolutionary conceptions have great relevance to this whole problem, which arises from the attempt to understand the relationship between different inherent attributes of men and other conscious beings.

When, for example, it is asked how one can know that one's fellow men (or animals) are conscious beings like oneself; or how one can know that when they stand with us in front of a pillar-box, the red that they see closely resembles the red that one sees oneself, it is surely relevant to take into account the principle of hereditary affinity between members of the same species, and to a modified extent, between members of related species. This seems reasonable, when there are no specific grounds for thinking that different hereditary factors are involved (as in the case of colour-blindness). Disbelief in the consciousness of machines may be partly grounded in the lack of hereditary and evolutionary affinity between me and my computer, if I use one. This may be inconclusive, but it is a relevant factor. It was much easier to believe that animals were automata in the absence of evolutionary beliefs. (This suggests one way in which evolutionary beliefs may have ethical significance.)

A solipsist could not admit having real parents and ancestors, but rather automata that miraculously generated and raised himself, the unique non-automaton. His 'children' could only be automata made from the coupling between an automaton and the non-automaton. A solipsist is obliged to deny the validity of hereditary principles, in a fundamental respect. In a God-like way, his consciousness would be self-created. Solipsism is absurd, and can be shown to be absurd.

The privacy of conscious experience, that has troubled philosophers and psychologists so much that they sometimes deny it, may have developed in the interests of individual survival and 'self-possession': interests that may set a different value on knowledge relating to oneself and knowledge relating to others. If so, the 'problem of other minds' may arise, at least in part, from the need of individuals for 'private and confidential' means of awareness of their own states and processes. If this makes good sense practically or adaptively, we should be resigned to any philosophical difficulties it engenders. To philosophize, it is necessary to stay alive. This topic is taken up again in the next chapter.

If one thinks of individual consciousness as distinguishing one individual from another, each with his own unique subjectivity and stream of experience, this cannot be derived from the individual's genetic uniqueness, since genetically identical twins enjoy this subjective uniqueness, feeling their own aches and pains, and are not themselves confused, as outsiders may be, as to 'which is which'. How can this be reconciled with their genetic identity, if this implies, for example, that barring developmental accidents they will see the redness of a pillar-box in the same subjective way? One falls back on the formula: however similar their perceptions, they are numerically distinct.

But whence comes this distinctness or distinctiveness of an individual's experience, if not from the specific hereditary endowment of each individual? In any profound sense, this may strain our understanding. At a more superficial level, one may reply that this distinctness constitutes one aspect of individuation and independence of function and action. Individuation, in its higher animal forms, goes along with powers of self-development and control over individual behaviour. The privacy of consciousness matches this power of individual control. Through an obsession with epistemology, at the expense of praxology, philosophers have tended to concentrate on the observer-concept of privacy, and to neglect the agent-concept: 'I did it myself on my own', 'I kicked out with my feet'. These are two adaptively correlated aspects of individuation. Only the individual that is burned feels the pain of the burn, and only he can respond self-defensively by moving away from the heat. Others may see that he is burning and move him away if he is unconscious, but moving another's body is as different from moving one's own as seeing another's hurt differs from feeling the pain. One's actions, like one's feelings, are uniquely one's own, and this own-ness is a basic form or aspect of individuation. One can no more act with another person's hands than one can see through his eyes. Those who deny any privacy for consciousness seem to be disowning and 'disactivating' themselves. They have passed beyond solipsism towards 'non-ipsism'.

The question I have been skirting can be put off no longer:

how is it conceivable that consciousness should develop from
unconscious precursors? This is relevant to both the ontogenetic
and the phylogenetic sequence. There seems absolutely no way
of determining precisely at what evolutionary stage individual
organisms became aware of their immediate environment or their
reactions, or how this change is to be understood. This is one of
the greatest gaps in evolutionary understanding, and it may never
be filled in or adequately resolved. Only those who believe that
the difference between a cabbage or an automaton and a sentient
individual is of small account, will minimize the significance of
this incomprehension.

Similar difficulties arise concerning the genesis of consciousness
for the individual, in the course of his development from the egg.

To escape these perplexities, the notion of omnisentience or
panpsychism has often been put forward. If all 'individuals' (in
a sense as elastic as you like) are conscious 'in due degree', the
question of the origin of consciousness is swallowed up by the
question of the origin of the universe (if it originated). In the
particular case of the development of a man, ego-consciousness
would be derived from egg-consciousness, which would be
derived from the consciousness of its gametes before fertilization,
which would derive from the consciousness of *their* precursors . . .
in an uninterrupted stream of consciousness from the original
universal source.

Well, why not? On the other hand, why? Theoretically,
panpsychism is vacuous, at least in the absence of any attempt
to show how the attribution of consciousness 'in due degree' to
any given class of individuals might help to explain their behaviour. Failing this, consciousness is degraded to the status of
cosmic shine or tinsel, an epiphenomenal grace-note, which it
surely is not in the life of man. (See next chapter.) This may
connect panpsychism with its extreme opposite, the total denial
of consciousness, when this is grounded in the belief that the
'hypothesis' of consciousness is functionally unnecessary and
empty and explains nothing.

It is doubtful, in any case, whether the question, how consciousness can be conceived to arise from non-conscious precursors,
can be escaped by the assumption of a continuous and universal
stream of consciousness. For this assumption is at variance with
the oft-experienced 'resurrection' of consciousness, in the life of a

man, after he has been anæsthetized. If the further assumption is made, that complete anæsthesia never occurs, this seems implausible and unsupported. This being so, it is unnecessary to consider further implications of panpsychism, such as the possibility that *post mortem* anæsthesia is not total either. (Perhaps this is the real 'unconscious' motive of the doctrine!)

In the development of a human being (or an animal, where applicable), the relationship between conscious experience and non-conscious processes and precursors often appears organized and adaptive. There can, for example, be no perceptual awareness of the environment without the unconscious development of the neuro-sensory system during embryogenesis. (If embryogenesis were a conscious process, and could later be recalled, this knowledge would have no useful transfer-value to the conditions of post-natal life.)

In later life, this relationship between conscious and non-conscious processes may be reversed. Unconscious habits may be developed by conscious effort and learning. The real if ill-defined limits of conscious attention make this kind of conversion necessary and valuable, as an aspect of mental economy. Another aspect of this is pain, that concentrates attention on particular bodily functions of which one is normally unaware, when there are more urgent affairs requiring one's attention. The selectivity of consciousness and the uses of unconsciousness are inherently adaptive, though subject like all other adaptations to 'error'. If one thinks in terms of an evolutionary growth of consciousness, the corollary or complementary principle is of an economizing and concentrating of consciousness. Psychic evolution is, in part, the history of this mutuality and complementarity between conscious and unconscious processes. The 'gulf' between them is an essential aspect of their interrelatedness, if it can be interpreted as serving the adaptive function of keeping 'in mind' only those tasks and projects needing concentrated control and attention. To concentrate is also to eliminate, or 'sub-liminate'.

Memory is an impressive example of this inherently adaptive and organized relationship. The unconscious retention of past experience, and its availability for selective conscious recall, facilitate the relevant concentration of our 'dispositional' knowledge of the past on whatever matters are of immediate or future concern, through the changing situations of life. Without this inborn power of selective retrospection, our consciousness would be jammed with a chaotic mass of memories. We are innately prepared or 'programmed' to retain and recall our experience selectively, this being necessary to the conduct of our lives, as an 'adaptation to the future'. We are innately predisposed to re-member the events of our future history as they become past, and to form our expectations and projects on this basis. There is *advance* provision of a *retrospective* faculty in the interests of *prospective* action. The time-shape of human life is thus preformed or pre-intimated.

In this intricate and 'ingenious' time-scheme, our conscious recollections and anticipations depend on the unconscious hoarding of past experience in an 'accessible' form: mnemic capacities which have been developed unconsciously through a genetic programme, which operates before the individual has any experiences to remember, in anticipation of his past, as it were, or of his future need to remember. There is a genetic *a priori* control over the time-structure of human experience. We are born time-surgeons, cutting our lives into past, present, and future, and connecting them up so that the interrogation of one's past becomes an intimation of one's future, and the dangers of being cut adrift from one's past are averted, as well as the opposite danger of being over-saturated in the past through an excess of indiscriminating past-consciousness.

The versatility and virtuosity of this unique 'arrangement' excite wonder.

Memory bears persuasive marks of a purposive origin: the interrelationship of conscious and unconscious processes, adap-tively regulated in the interests of mental concentration and coherence. (This is, so to speak, the 'good' unconscious of reten-tion, not the 'bad' unconscious of repression.) The selectivity of recall, by no means infallible, is surprisingly apt at picking out the detail of a unique occasion from the past. Most impressive of all is the regulation of time-relationships, both as subjectively experienced and prior to experience, in that the ability to retain and recall previous experience, and to form expectations, is genetically programmed in advance of experience. Our passage through time has been pre-charted.

Memory is a marvellous adaptation to time and the transience of experience, which it partly overcomes and transforms into 'transcience', or knowledge of the transient. It is truly hard to comprehend how it could have come into being through an evolutionary process that is a 'pure process' or blind sequence of changes, unguided by any sense of relationship between earlier and later changes.

5 Pain

Pain is a great problem, both for purposive and non-purposive philosophies of life. Its ambivalence, or 'ambivaluence', has various aspects, one of which is this: that the value of pain derives from its disvalue, from its unacceptability. If suffering were not insufferable, we should all be quickly dead, from apathy. To be insufferable is its *raison d'être*. The painfulness of pain is a life-saving tautology. Its intrinsic experienced harshness makes it uniquely apt for its protective and admonitory functions. It may be a fallible monitor, but it is an indispensable one. Under this harsh protection, pre-ordained, are our lives drawn out. A small hurt may prevent a greater harm, and a premature death. The purpose of pain is served by its cessation, through the defensive reactions it urgently provokes. Pain is an evil we cannot live without, and should not seek to live with—unless we are bent on suicide, or are incurably afflicted. To be heroically tolerant of pain defeats its curative or aversive purpose, which would be endangered by ease of habituation; and when circumstances make long tolerance necessary, such resignation is a most unnatural reversal of the healthy response to pain, that of 'impatience', and may be impossible unless some other purpose can be substituted. No one would seek these other purposes and reasons—for putting up with pain, rather than putting it down—if the primary purpose were always realizable. This is another aspect of pain's ambivalence. Man is the only being on earth that seeks a reason for suffering, and is appalled when he cannot find one. This adds to his suffering. On the other hand, only man can take the quick way out of irremediable suffering. Yet for the great majority of men, suffering is a lesser evil than death. It is as though the life-saving purpose of pain were too strongly entrenched for easy conversion into its opposite, even when this purpose fails.

Suffering serves life, not death. One may die painlessly, one cannot live painlessly.

The most sweeping denial that pain has any purpose at all (humanly and animally speaking) was implicit in the nineteenth-century idea of Epiphenomenalism (which in this context may be called 'Epipenalism'). This is nihilism with a vengeance. If pain is inherently gratuitous, and totally without effect on behaviour, we may well believe in the existence of an evil and sadistic Creator. Amazingly, the idea seems to have been formulated with no real awareness of its terrifying implications. If pain has no admonitory function, why should it be generally associated with situations of danger and imprudent acts? Innate insensitivity to pain is a defect that can prove fatal. We may suffer too little (or too late) for our own good. There is no reason to think this the worst possible world, made on the principle of pain for pain's sake. Epiphenomenalism serves to remind us that things might be a lot worse. Its uncalled-for nihilism demonstrates that anti-purposive philosophies have their excesses and absurdities, no less than the pan-teleology of former times. It is as easy, and futile, to pick a quarrel with the cosmos and invent grievances, as to fabricate favours and harmonies.

Behaviourism also, in any form that denies the privacy of pain as experienced, makes nonsense of its purposive functions. These require that every individual should have his own private monitor or alarm, to which, as an independent agent, keen on surviving, he can uniquely react. How absurd it would be if the individual's pains were public property or communal experiences! It is precisely the privacy of pain that gives it its value as the protector of the individual (or, in the case of pregnancy, the mother-and-child unit). This privacy is the very reasonable correlate of individuality. The solitude of pain reflects the self-defensive capacities of individuals. Private is as private does. In the doctrine of Nirvana, suffering ceases when individual separateness is transcended. In its protective function, pain is the preventive awareness of imminent personal danger, which stridently forces attention and tends to dominate or drown all other thoughts and perceptions. First things first.

Pain does tend to isolate the individual and cut him off from everything outside himself. At moments of danger, the individual has to react on his own, in his own defence, instantaneously. Any individual is always, and uniquely, 'at hand' for himself. If asleep, pain may arouse him. If he cannot help himself summarily, probably no one else can.

If we think of pain as having an evolutionary history, its origins must go back beyond the possibility of social aid for individuals at risk, and before the development of public expressions of distress. The evolutionary 'problem' was to supplement private distress signals with public ones, when these were capable of arousing a social response. The expressiveness of the face as a social screen, open to all except one, is a good example; its social significance derives from its being normally naked (physically and expressively) and invisible to its owner, whose feelings are expressed blindly. No one needs to scan his own face to know his feelings.

The distinction between private awareness of feelings and their public expression has a vital part to play in human life, a part that is, in general form and possibilities, predetermined by our nature as sociable individuals. If, as a general rule, self-awareness outstrips the social expression of feelings, in range, subtlety, accuracy, and immediacy, this may reflect the more recent evolutionary importance of social expression, with a continuing bias in favour of individual independence and self-reliance. This has wider ramifications than can be pursued here.

The distinction between pain and its public expression is appropriate to a situation where people may expect help from others, but cannot rely exclusively on it. Privacy follows the principle of individualism and separateness, publicity the principle of cooperation and compassion. There is I-knowledge and there is eye-knowledge of suffering, and to confuse the two is to confuse who's who, who the victim and who the helper, who the patient and who the doctor. People who feel sick when they witness sickness need help, rather than give it. To relieve another's agony, it is essential that one should not be agonized oneself. On the other hand, if one did not know what it was to be agonized, one's sympathies would be checked. The experience of pain has a dual role—as private monitor, and as the basis of compassion. In reverse, the awareness of another's suffering, as well as evoking aid, may serve as warning. Man, uniquely, can learn from the sufferings of others, as well as his own. On this is based the possible deterrent effect of punishment, on those not punished. It can lead to exploitation, when some are made to risk suffering that others may escape it. Such risk-taking was the allotted role of the taster of royal dishes, in certain courts. The most prevalent instance today is the experimental induction of disease in animals, *pro bono humano*.

The heightened awareness of the sufferings of others has its own ambivalence or ambivaluence. Its dark side is that without it torture would be impossible. Similarly, without a self-conscious attentiveness to pain, self-mutilation as practised by masochists would not exist. But if these had been the dominant responses to pain, and the social expression of pain, man would have destroyed himself long ago. (Pain becomes a total 'disvalue' for survival when it is valued for its own sake. It is only valuable when disvalued.)

It is hard to hurt a person badly without harming or injuring him; this testifies to the admonitory value of pain. The torturer's 'problem'—how to keep the tortured alive, without abating the pain.

From an opposite viewpoint, it is easier today than ever before to relieve pain and leave the vital disorder untouched. Two of the goals of medicine—the restoration of health and the relief of suffering—are becoming increasingly independent. While effective pain-suppressors are an enormous boon, the danger of their availability is that the admonitory functions of pain will be reduced. Pill-taking is a bad substitute for treatment, when treatment is possible. Subjective dis-ease is still an indispensable pointer to bodily disorder (though these are more independent than is implied by their etymological conflation in the word 'disease'). The 'paternalistic' premise of pain is that people cannot be relied on to act prudently as regards their self-care and health, unless rapped and racked when danger threatens. Preventive care may diminish, but can never replace the need for pungent warning. A vulnerable fortress requires alarm systems at its weakest points. The enemies most to be feared are those that elude these detectors, and reveal themselves only when victory is certain. If such enemies were frequent, the fortress would not stand for a day.

Pain supplements the unconscious defences of the body and formation of antibodies, through its effect on the behaviour of the whole organism, and its symptomatic value for diagnosis.

There is much, then, to suggest a purposive origin and evolution of pain and the reactions to pain: its subjective privacy and primacy or intensity, so apt to its individual warning functions; the overcoming of the asocial, solipsistic implications of this privacy through the expressiveness of the human body and voice (the piercing cry, the plaintive groan, the far-moving scream), appealing to the sympathetic response of others who are not inhibited by personal involvement in the sufferer's situation;[1] the general (but not universal) tendency for the severity of pain to reflect the acuteness of the threat to the health and life of the pained one; the built-in protection and reminder that pain affords against the vulnerability and fallibility of organisms, with respect to their inner needs and unseen functions, and their overt behaviour; the 'flighty' or vanishing streak in pain, when it generates action that brings about its own decay and decease—the self-destructive 'intent' of pain. The introduction of a state of consciousness whose function is to be self-eliminating, through the correction of associated 'errors' and malfunctions, is a very subtle stroke for the self-protection of organisms. To think of it as the fruit of unconscious mechanisms of evolution strains belief. Of such mechanisms, one may say disbelievingly:

> They that have power to hurt and yet feel none,
> That do not know the pains they most do sow,
> That moving others, are themselves as stone,
> Unmoved, cold to tribulation's woe.

Could these stony mechanisms have originated the sufferings of men and animals, and the possibility of their relief through the implantation of compassionate feelings and dispositions? This is the Medusa myth reversed. (She turned men into stone.) The miracles of materialism include that of the Unmoved Movers, the unfeeling specifiers of appropriate feelings, that 'prescribe' pain for others.

The notion of a private, individual experience, with an intrinsic quality well suited to its role of ill suiting its sufferer, that has to be experienced to be relieved—for the imperative need for relief to be apparent—the notion of such an experience having been developed and prescribed for organisms under certain conditions of danger, during the course of evolution, by antecedent causes

[1] The cries of animals in pain seem of doubtful utility (except when human help is available). Perhaps they can function as social danger signals.

and mechanisms, is one that borders on self-incoherence and contradiction. If the private experience has been prespecified for its intrinsic or innate admonitory aptness, from the viewpoint (or feeling-point) of a conscious subject, how can it be regarded as private? Is not this rather like speaking of the outside source or origin of a secret confined to an inner circle? There is something very paradoxical in the notion of the privacy of pain, its pain-fulness-for-the-individual, having been prescribed for him under certain circumstances, and not originated by him. (For most of us, pain is nearly always involuntary and unsought.) In some sense, surely, this privacy must have been breached or transcended 'at source'—in the original specification of pain-sensitivity in animals and men, during their evolution. How else could it have been imposed on them, made inherent in their make-up, as an individual affliction and adaptation? How is suffering possible? Was it originally thrown at us all, men and animals, by molecular malice or indifference?

If there is a creative purpose informing evolution, the paradox of a private experience, significant only from the viewpoint of a conscious subject, being predetermined for the individual by factors outside his consciousness and control, is diminished. If individuation, with all the separateness that implies, has been pre-purposed, there can be no secrets for such a source. (This would be the ultimate source of 'secrets' and 'secretiveness', arising from the separateness of conscious individuals.) The significance of pain for a conscious agent would have been pre-established by a creative act that transcends all merely human categories of 'private' and 'public'. Concerning the relationship between the experience of pain and the relevant properties of the nervous system, the former could be characterized as the inherent, pre-established purpose, goal, or 'meaning' of the latter (and a similar interpretation could be given to many other relations between conscious states and their underlying figuration).

If sensitivity to pain has been developed purposively, as a private and individual warning signal, the overcoming of the drawbacks of this privacy in a social species by the development of public expression, social cooperation, and compassion, could be thought to proceed from the same source as originally established the private functions of pain—and in establishing its privacy, for all normally constituted individuals, transcended it. To originate and impose pain, as a monitor, without feeling it, as we feel it, might be compared, however remotely, with a fantastic feat of the creative imagination. This is surely not gratuitously speculative: there is no coherent 'scientific' account of the origin of pain that is being irrationally discarded or disregarded. The great problem is, how can suffering be understood as predestined for the individual, and its role predetermined, by unconscious unempathic processes and mechanisms? Is not pain too bad for materialism to be true? Too bad to have been self-imposed on organisms through the ministrations of their own genetic systems, as these evolved painwards? This, of course, constitutes no kind of 'proof' that the purposive interpretation is correct; and there are familiar objections to it, which are strongly and bitterly felt.

The apparent perversity of ascribing a purposive origin to pain lies in its foul association with bodily disorders and deficiencies, errors of behaviour and follies of negligence and intemperance, convulsions of nature, famine, plague, pestilence, fire . . . . But this is to deny the purposiveness of a warning because of the evils against which it warns. The purpose of warning depends on there being something to warn against, and cannot possibly be invalidated by the existence of the latter. The ambivalence here lies in the fact that no warnings would be needed if everything was perfectly adapted and submissive to the ease and interests of living beings. But this cannot, in the present context, affect the issue of a limited purposiveness. Anyone who thinks that, if suffering is the necessary price for the survival of higher conscious beings in such a world, it would have been better if they had never been born, should recommend universal euthanasia, the instant destruction of human and most animal life, by bomb or poisonous gases.

More relevant and less utopian criticisms may be directed at deficiencies of the warning system, errors in the error-correcting service. Absurdly disproportionate and disabling pain may be felt over trivial disorders, or none at all. Conversely, when warning is most needed, it may be wholly absent. Such inefficiencies may be compared with those of sense-perception, which also has its blind spots, inaudibilities, and illusions. These do not seriously detract from the normal serviceability of perception. Errors of the pain-monitoring service seem more grave and scandalous, precisely because they are painful errors, and quite often a matter of life or death. They cannot be regarded with academic or æsthetic detachment. No one minds psychologists inducing optical illusions as an experiment, but experiments on pain are different. There is, however, an inherent fallibility in all information-services, and those concerned with pain are not exceptional in this. To a certain extent, errors can be guarded against, and their impact lessened. The vital 'premise' of pain is that errors that are lit up will often be corrigible. If we develop awareness of characteristic inherent errors in this error-monitoring service, we may be able to compensate, by our intelligence, for this insentience. This, in turn, carries the risk of errors of the third degree. There are no infallible monitors. It is necessary that some men should die unnecessarily early, when a warning might have saved them, however mediated.

Even when the pain-monitoring service works well, by whatever criteria are thought appropriate, there may be little or no chance of a corrective or remedial response to the danger to life and health. The dis-ease may be capable of relief, the disease or deficiency or injury may be incapable of remedy. The 'warning' is then no true warning, being a warning against the inevitable, or the registration of a *faute accomplie*. This misfortune seems inherent in any general, prearranged alarm system, in an uncertain world, in which the chances of an individual, with or without outside aid, being able to respond effectively to threats (from within or without), depend on his individual circumstances and the contingencies of time and place. A man or animal tormented by thirst is driven to seek water, whether it be within reach or not. The warning that saves one, does not save another (who would have perished anyway). In a sense the unfortunate ones suffer for the general good (which might, with better luck, have been their good). A limited concept of purposiveness is not invalidated by the constant risk of fruitless suffering and irremediable affliction—a risk often increased by a refusal to heed early warning signs. Pains without remedy are the dark side of a situation in which remedies without pains are improbable and unsought, for many bodily ills. Since death is certain, some diseases are certain to be irremediable or 'terminal'. (Or vice versa.) To 'indict' pain for this is like blaming the barometer when it indicates the imminence of a tornado. It is the permanent possibility of death that gives pain its uses, and the final certainty of death that sets its limits. Immortals would have no use for it. Only death-resisting mortals can use it, by opposing it for as long as possible. Everyone knows, at least 'in theory', that resistance to death is finally hopeless, yet resistance hopefully goes on. The struggle against death is probably unique, in this. It is not very 'rational'. It is a prerational struggle that we are committed to. Man's unique insight into the final futility of the struggle has made amazingly little impact here, by and large. He uses this knowledge of death more to reinforce the struggle than to despair of it. When he risks his life in wars and other enterprises, it is very doubtful if despair over his mortality plays any significant part in inducing such recklessness.

Perhaps the worst kind of irremediable suffering is that associated with 'chronic' ailments, which may not be especially dangerous but are both incapacitating and distressing over long periods, with little hope of remission. The suffering gives urgency to the need for more effective treatment of the ailment, but there is no guarantee that this will be found. This applies to all kinds of chronic illness, whether categorized (uneasily) as physical or mental.

The interpretation of pain as a valid warning is harassed and limited by hard cases and cruel exceptions. One may still maintain that the prevailing (and predetermined) use of suffering is the corrective, admonitory one. No other general interpretation seems to fit as well, despite the grievous limitations.

A radical criticism of this interpretation is that it is no more than a figure of speech; that it uses an over-rational and under-severe metaphor derived from a human practice that is more prescient and more humane than pain. If pain be a warning, it may seem a worse evil than that which it warns against: a 'warning' we would wish to be forewarned against, so as to forestall. A warning is usually a verbal communication that is painless, though intended to disturb, and to prevent the warning from coming true. When the term is used figuratively, other than in the context of human communication it is clear that the warning is read into the situation (or out of it) by a human interpretation or projection. 'A red sky at morning is the shepherd's warning.' Similarly with the saying 'Let that be a warning to you.' The warning functions of pain are man-made, the fruit of man's verbalizing, rationalizing, projecting abilities and his sense of time. Pain 'in itself' is no more than an organic stimulus tending to excite a self-protective reaction of an immediate, unreflective, and unhypothetical kind. Only for man can this stimulus become the ground for a prognosis (in conjunction with other symptoms). It would be as absurd to think that the 'natural function' of pain is to warn, as to think this the natural function of red skies and traffic lights.

En passant, one may note that 'stimulus' is a Latin word meaning 'goad', a spiky club used for driving cattle. While this adds point to the use of the term in the present context, an important difference is that pain, unlike a club, is a private goad or stimulus, not something that can be switched from back to back. The club is not the pain. Clubs are man-made, pains can only be inflicted on beings that are painable 'by nature'. Clubbing the furniture is innocuous, a harmless sublimation of rage and cruelty.

One man's goad may be another man's warning. Self-protective responses to pain vary considerably in immediacy and prescience.

The warning-metaphor, like the stimulus-metaphor, is one that has some misleading associations. But even when there is no conscious apprehension of specific dangers to be warded off, pain does alert the individual and often permits a defensive response before serious harm is done. When the intensity of the pain and the acuteness or closeness of the danger increase proportionately, by a gradual escalation, the comparison with an early-warning system seems apt. The fully conscious and explicit realization of future dangers, and the intelligent appreciation of warning signs, may be seen as a rational development of the primitive protective functions of pain; rather as the conscious aim of begetting a child may be seen as a development of the procreative instinct. (The 'pains' of childlessness might also be regarded for their protective function—for the species.) Pain may fail in its protective function, as sexuality may fail in its procreative function. The latter is the more bearable failure. The preservation of the race is not so critical an affair as the safety of the individual, is not put in hazard by a single failure. And we are more reconciled to unproductive rapture than to unproductive pain.

The odious superstition that pain is punishment has been aided, presumably, by the connection between sexual licentiousness and disease. The absurdity of this is that wholly 'innocent' people are infected and the 'guilty' often escape.

The 'enlightened' attitude to pain and its *raison d'être* seems greatly superior to that of the pain-moralizers, spiritualizers, and obscurifiers, despite its rationalizing failures and its apparent inability to explain the genesis of pain. The alleged connection between pain and 'profundity' may sometimes hold—as in the slow movement of Beethoven's late A minor quartet, *Heiliger Dankgesang eines Genesenen an die Gottheit*; but what may especially move one here is the contrast between the desolate, immobile sections, marked *molto adagio*, and the upsurge of returning strength. (Schoenberg similarly treated an illness and return from death—his heart stopped beating—in his string trio, an extraordinary work.)

The enigmatic aspect of pain is that its 'raw feel' is manifestly not the result of human or animal choice, yet seems incredibly well chosen for its significance from the viewpoint of a conscious being bent on living for as long as possible in a dangerous world. If you throw a stick or plant in the fire, it will lie there. Incapable of being 'moved' by their burns, they neither suffer nor survive. A sentient being may suffer without surviving; at least he has a better chance of surviving. Suffering has not been inflicted on beings that constitutionally are totally incapable of profiting from it. Indirectly, however, such beings may be 'protected' by pain— through the protection this affords to the sentient beings that care for them.

6 Love

With respect to its biological function, pain may be compared, as well as contrasted, with the love between people of opposite sex. The differences are more obvious than the resemblances. Pain, in its egotistical function, demands one's self-attention, whereas love goes out to another conjunctively. *Doleo, ergo sum; copulo, ergo sumus.* The most important resemblance perhaps is that pain and love are highly distinctive experiences which are, as it were, forecast through the innate constitution of the individual as pre-established possibilities. Each in its different way has a conserving function, pain being prescribed for the protection of the individual and sexual love for the conservation of the race. The 'premise' of pain is that premature death may be avoided, the 'premise' of sexuality is that death is inevitable. Functionally speaking, pain is death-resisting, death-deferring; love (in its sexual aspect) is death-accepting. In the midst of love we are in the 'field' of death. Granted this interrelationship, the cost of not dying would be not loving sexually. To value the latter, while deploring its condition (mortality), is unrealistic. Not all goods, or imagined goods, are co-possible. Love and immortality do not mix. As for non-sexual forms of love, at least some of these may be thought to depend in some way on Eros (and hence on death as its condition), or to be substitutes or sublimations of Eros.[1]

If pain be the necessary condition of our (limited) powers of survival, love is the greatest positive asset and associate of our mortality. Here is concentrated life's irresoluble 'ambivaluence', in the deep organic linkage between value-opposed states and experiences; between Eros and Thanatos, between suffering and surviving.

[1] The love-death connection can be extended by analogy to cultural processes. The love of novelty and creative zest feed on the conviction that all vitality has departed from defunct styles and idioms. Belief in the 'eternal' or 'immortal' status of the cultural inheritance makes for sterility.

If one asks what it is that distinguishes human love from animal mating, no simple answer is possible. Part of the answer is the awareness that all love is mortal, which casts a certain tension on all loving, a future cloud on a relationship no longer enacted in a mindless, timeless void. It is only in the sexual embrace that man may recover this lost state, temporarily. Its ecstatic quality is inseparable from this lost sense of time, care, and reflection. The paradox, if it is one, is that an act so fraught with possible future consequences should have this subjective timelessness. Ecstasy is a 'standing-out' from the temporal and causal modes of apprehension. Almost by definition, it is a short-lived and exceptional state, and cannot sustain, though it may support, any lasting human relationship.

It may appear irrelevant, in the present context, to mention non-personal forms of love, such as love of knowledge, as distinguishing the human from the animal world. But is not this link between love and knowledge present also in the love between persons, distinguishing this from animal relationships? A lover wishes to 'know' his or her beloved as fully as possible, and to commune with him or her in ways that are simply not open to animals. Man's unique amative-cognitive capacities are not unrelated. The odd biblical phrase 'carnal knowledge' implies that even on this level there is a getting to know the ecstatic potentialities of bodies joined together, which like all knowledge may be put to some use.

The relation between love and knowledge is too complex to follow up here. But the suggestion may be thrown out that, motives of utility apart, the drive to know usually derives from love or interest for the other person—or for the impersonal process or object. This drive may be disinterested but not uninterested.

Another facet of the uniqueness of human love is the sense of its uniqueness, in each of its 'occurrences': the sense of a unique relationship never repeated in the entire history of mankind, past and future. It is this sense of uniqueness that gives unrequited love its torment, and requited love its fears. Even 'monogamous' animals are incapable of feeling 'There's no other bird in the world like you.' Nor are they liable to 'infatuation', to the imputation of qualities to the beloved that are not 'there', or to falling in love with an imaginary being.

What may be called 'the idealization of love', the quest for perfect mutuality and unimpeded union and communion, is merely one of the possible variations that distinguish human and animal mating, and generate peculiarly human complexities of feeling and appraisal. But if animals could love in this way, they would cease to be animals, i.e., beings without ideals, and incapable of disillusionment. Love is rooted in man's animality, but expands beyond it. To identify it with sexuality is rather like equating patriotism with defence of territory. Sexuality itself, when humanized, becomes not merely a condition of reproduction, but also (for instance) a way of thinking about the world, in the grammatical category of gender. As sexuality got spoken of, so speech became sexualized.[1]

The evolution of love, in its sexual and wider meanings, challenges human understanding. Somehow or other the task of self-replacement has been made a thing of joy, oblivious of its death-implication. This 'duplicity' of love, whereby the intensest subjective desires are linked 'objectively' with individual mortality and species-survival, makes it unique among the passions, and makes the strongest impression of a purposively organized relationship between a complex of acts and feelings and inexorable future necessities. To suggest that the concordance between individual ecstasy and the interests of the species came about by chance, is very implausible.

Of course, natural selection or sexual selection will be invoked. Thus, feminine beauty is ascribed to the selection by past generations of males, of specially favoured partners, whose genes will have influenced the appearance of any daughters they may have. On this view, feminine beauty is the selective creation of man. To say that men love women for their beauty (*inter alia*) is a half-truth. The other half is that women are beautiful because they, or their female ancestors, have been loved by men, selectively. Women are beautiful because ugliness sterilizes—or rather, reduces relative fertility (or fertilizability).

This seems fairly incredible, though conceivably it might be so. But consider another example, the beautiful plumage of male birds, such as the peacock. To imagine that his splendour has been brought into being by the sexual preference of generations of

---

[1] The sexual and organic metaphors of grammar, such as conjugate, copula, generative, stem, root, emphasize the analogies between languages and living beings.

selective females is to assume that the æsthetic taste of these has predominantly favoured those particular combinations of colours and decorative patterns, from among the chance variants thrown up by genetic mutation and recombination. It is not easy to understand how this similarity of taste could have developed, or what biologically useful function it could serve. But even if this were explicable, there would remain the astonishing convergence between the sexual preference of birds and the æsthetic admiration of men. When the plumage of male birds or the beauty of wild flowers is ascribed to selection on the part of female birds and of insects, the implied æsthetic coincidence between such remote forms of life as men and birds, or men and insects, is no less inexplicable than the original phenomenon of natural beauty seemed to be, before the selective explanation was thought up. Of course, not all the phenomena that are now generally ascribed to sexual selection arouse the æsthetic admiration of men. But a high proportion do, and to say this is purely coincidental seems feeble and over-confident.

Anyone who claims to know how on this planet love originated and developed is, in my opinion, over-confident. As an innate human possibility, love appears to have been genetically pre-scribed for human propagation and the establishment of families and rearing of children; yet its subjective, interpersonal signifi-cances are only realizable 'immanently', from within the relation-ship as lived and experienced. How is the reality of this emotion attested? *Amando. Docet experientia.* But in their innate possibility, these unique experiences originate from beyond the circle of experients and prior to experience, in whatever 'factors' have determined these human possibilities. *Mirabile dictu*, these compel-ling and esoteric experiences have an extraneous source or genesis. To deny this would imply that love was a complete invention (or convention) of man's. I am sure I did not invent it, nor I think did any other man or group of men; which is not to deny that love may be inventive, and manifest itself in many variable ways. But the basic potentialities are inherent in man. Man, so to speak, proposes, God (or Nature) predisposes.

If love is not a human invention, is it then an evolutionary invention of germ-cells (through mutation) in the interests of their own fusion or fertilization and the development of family life, with a little happiness thrown in as a bonus? I find this incomprehensible, a mystical shoot from the materialistic philosophy and divinization of DNA. That so bizarre an idea should have come to be taken for granted demonstrates the power of the *Zeitgeist* and a scientific philosophy. The origin of love seems to me a scarcely penetrable mystery. By comparison with this occult origin, much of our knowledge and source-knowledge may seem somewhat peripheral and arid; theory at its greyest. The endless psychologizing that goes on round this 'subject' evades this mind-straining question of its primary genesis. We might understand our life situation better if we better understood our lack of understanding on certain fundamentals.

Is this mere sentimentalism, a swooning over love's mystery? No, surely the mystery is a genuine one: how human values are related to the non-human world-process.

That love should be prescribed is a good antidote to the near-consensus of philosophers today, that the universe is quite alien and detached from all human intimations of value. Such philosophies really ought to take more seriously the question of the origin and possibility of human values in such a universe. If these values are not wholly created by man, but depend on his innate predispositions, the question of their non-human source cannot be evaded. Any world-process that has assisted in the establishment of love as a prime value of life cannot, surely, be fully intelligible in physico-chemical terms. I cannot think that a non-religious view of life can be finally satisfying or adequate.

Though there may be a chance element in the possibility of love arising between individuals, this does not imply that love as a possibility and propensity has come into being chancily. If one is invited to a feast, it may be an accident whom one sits close to; but the feast and the invitation are not accidents.

Love leads to the renewal of the species which leads to the renewal of love. *Cras amet qui numquam amavit.* To the unknown amorific agency, praise.

7 Values

There is, I believe, a problem of surplus value that is quite analogous, from an opposing viewpoint, to the theological Problem of Evil. I wish to suggest that scientific materialism and theology are faced with inverse quandaries that resemble one another sufficiently to evoke similar attempts at overcoming their respective problems, in this domain of value. The escape from the problem of unnecessary evil *via* atheism may lead straight into the problem of unnecessary good.

The basic issue is whether the human conviction that certain forms of activity and experience, of richly varied quality, are intrinsically valuable in themselves, can be plausibly reconciled with the notion that man has evolved from the animals by a value-neutral process of evolution, oriented only by survival 'value'. It would probably be advisable to drop this latter use of the term 'value' and reserve it for situations that involve some element of conscious evaluation and discriminating choice. Survival is not so much a value as a condition of value, of valuable acts and experiences. If survival *per se* be accounted a value, it is difficult to see by what criterion such excellent survivors as planets and galaxies could be excluded from the domain of value. It was not on this basis that value was previously attributed to planetary existence, but on the basis of the Harmony of the Spheres, the cosmic symphony of stars.

We men at any rate are known to attach enormous importance and value to various interests, experiences, and activities which are not linked in any obvious way to our ability to survive and have progeny. By the criterion of biological utility, such values appear surplus. It might be said that they make life worth living and prevent apathy and accidie, enemies of mental health. It is possible, for instance, to praise the use of music for purpose of soul hygiene. But if so, it remains mysterious why our mental health should be dependent on non-utilitarian values. The concept of health is not value-neutral.

Values that are biologically surplus present a kind of mirror problem in relation to the theological problem of unnecessary evil.

A common theological response to the latter is: Man is endowed with free will. He can disobey. A similar reply may be given, from the viewpoint of scientific materialism, to the problem of man's biologically surplus interests and values. He can flout the utilitarian programme of organic evolution which created him. He may, for example, take special measures to prolong the lives of individuals who would otherwise be speedily eliminated in the course of natural selection.

A possible materialistic defence here is to say that, for reasons of survival, man has been endowed by natural selection with an intelligence and imagination which greatly expand his freedom of choice and powers of discrimination, and make him much less dependent, as an individual, on instinctive performances and preferences, and as an evolving social group, on the processes of natural selection. This independence and imaginative liberty gives man far greater adaptability to different conditions of life, and also makes it possible for him to conceive and follow some non-utilitarian values and interests, as a kind of bonus or by-product of the adaptive value of man's imaginative powers. As evil, on the theological argument, is the unintended by-product of the gift of freedom, so surplus value, on the materialist argument, is the non-selected by-product of man's adaptive imagination and independence. This biological thesis might be seen as a Christian or post-Christian heresy.

Neither the theological argument, nor its materialistic mirror or inversion, seems adequate for their respective purposes. Granted that man has developed a certain limited independence of biological needs and freedom from the rules of organic evolution as these have hitherto prevailed, this will not by itself explain such manifestations of this independence as artistic creation and philosophical understanding, or the immense value that a civilized society attaches to these. It remains a paradox that man could have achieved some independence of utilitarian ends by means of qualities developed by a utilitarian process of evolution. The 'emergence' of human values from a value-neutral process is every bit as mysterious as the emergence of evil from a Source presumed to be perfect.

Theologians would like man alone to be responsible for evil, as materialists or humanists inversely would like man alone to be responsible for all that is of pre-eminent value. But both have somehow to allow for the probability of man having innate predispositions, not of his own choice or making, which strongly incline him to evil deeds, on the one hand, and work of non-utilitarian value, on the other. Original sinfulness needs complementing by a no less 'original' or deep-rooted impulse to the perfection of work of enduring value.

Just as there exist evils not attributable wholly to freely chosen deeds, so there exist goods not so attributable. As in art the existence of *objets trouvés* is admitted, so in life elements of value are encountered rather than made. In certain respects the universe is not merely permissive of human values, but provides conditions favourable for their realization, at least in part. Human creativeness may be greatly stimulated by the appearance of natural forms, and the human desire to understand the universe is at least partially assisted by its inherent order, relative to human powers of comprehension.

The value that men attach to knowledge is one of the themes discussed by Jacques Monod, in his book *Chance and Necessity*. He emphasizes that this value is constituted by an act of free creative choice.

> In the ethic of knowledge *it is the ethical choice of a primary value that is the foundation.* . . . The ethic of knowledge does not impose itself on man; *on the contrary, it is he who imposes it on himself.* . . . No system of values can claim to constitute a true ethic unless it proposes an ideal transcending the individual self to the point even of justifying self-sacrifice, if need be.
>
> By the very loftiness of its ambition the ethic of knowledge might perhaps satisfy this craving for something higher. It puts forward a transcendent value, true knowledge, not for the use of man, but for man to serve from deliberate and conscious choice. At the same time it is also a humanist ethic, for it respects man as the creator and repository of that transcendence.[1]

Monod is here apparently denying the Aristotelian view that the

---

[1] *Chance and Necessity* (London, Collins, 1972), pp. 164–5.

desire for knowledge is inherent in man's nature. One is left wondering why men should be so keen to serve these values of their own creation, even to the point of self-sacrifice. Monod suggests that men may have a 'need' to transcend their individual selves. This seems to reintroduce the notion of an inherent 'want', of an idealistic or even religious stamp, which the theory of the free creation of values is intended to eliminate. The question therefore remains, how a value-neutral process of evolution could have endowed man with an inherent need for transcending himself through dedication to ideals that are not 'of use' to him.

Even if one were to accept Monod's contention that values are created by a free choice, with all the anarchy and arbitrariness this implies, it would still remain a mystery how man came to acquire this unique creative freedom, this independence of the evolutionary process to which he owes his being. There seems to be a dilemma here, for the Neo-Darwinian. If he says, 'Man is free to create his own values,' this freedom is something falling outside his categories of explanation; while if he says that man's values are partly rooted in his nature, or his need for self-transcendence, he has difficulty in explaining how such non-utilitarian needs have arisen. Faced by this dilemma, Monod seems to hesitate and take over inconsistent elements of both alternatives. It is difficult to understand how the conception of man's absolute freedom in the creation of values can be reconciled with a materialistic account of man's evolution.

Another of man's most striking and valuable propensities is his talent for humour and love of clowning. This is seldom noticed in discussions of evolution. It would be easy to expatiate, in a speculative way, on the contribution it has made to human survival, and presumably is still making: how it eases and sweetens social intercourse, relaxes tensions, disarms men's fears and aggressiveness (though some humour is aggressive), assists in the rearing of children, whose jokiness may be endearing, assists learning, prevents depression, and so on. Humour and sociability certainly go together. To weep alone is normal, to laugh alone abnormal and slightly manic. Humour is generally reassuring, unless directed at us or satirical. The world appears in a friendly, festive guise. The 'archetypal' human jokes and ribaldry, for instance about sex, may help to break down anxieties and inhibitions, and so promote procreation.

An important area where we need all the reassurance we can get is that of death. Perhaps gallows humour, in which jest and terror tag one another, helps us to face the unfaceable with derision, and to conquer the depression and paralysis experienced by those unable either to forget death or come to terms with it. 'If you can't beat it, ridicule it.' This may be especially necessary for people who have a daily brush with death, such as medical students and grave-diggers. The grave-diggers in *Hamlet* sing at their work and make grave jokes. 'What is he that builds stronger than either the mason, the shipwright, or the carpenter? The gallows-maker; for that frame outlives a thousand tenants.'

One might imagine that man became merry when he discovered death and his impotence, that this was his unique way of adapting to the knowledge that he alone possesses. But I suspect that death-mocking has a more recent and more restricted history than such an origin would imply. Death-merriment is dangerously unstable, and easily turns into contempt for life as a meaningless farce. 'The last act is bloody, however charming the rest of the comedy. A little earth is thrown on one's head, and that's that' (Pascal). This reminds one again of the grave-scene in *Hamlet*: this character that was once a great landowner now has his fine pate full of fine dirt. 'Imperious Cæsar, dead and turned to clay, Might stop a hole to keep the wind away.' Yorick, king's jester, is now quite chap-fallen, and finds no one to mock his grinning. 'Now get you to my lady's chamber and tell her, let her paint an inch thick, to this favour she must come. Make her laugh at that.'

No, this does not call for laughter—but nor does it call for excess of horror. To this favour I shall not come, since very wisely I shall have vanished much earlier in the proceedings. The sense of identification with one's left-overs is spurious. The skeleton has no individual or personal significance, but only generic, as an anatomical specimen. Hamlet had no way of identifying the skull he held as Yorick's skull. For this kind of reason the skull has acquired representative significance, as a symbol for the universal and impersonal attributes of death.

Hamlet is so strongly drawn towards corpses that he treats them like fellow-men, and talks to them.

From this brief discussion, it appears doubtful that humour has any special adaptive significance as an antidote to man's fear of death. Possibly humour has a more general adaptive

4

utility and origin, through its reduction of the tensions and anxieties of men's social and personal existence?

Possibly. But how uncertain, how apparently untestable, such functional explanations appear! Belief in the utilitarian or selective significance of humour may not be compatible with its continuance. Knowing that we have to jest, or risk possible extinction as a species, would make spontaneous laughter very difficult. If commanded to laugh (sincerely) or die, most of us would have to die. (We might, however, manage a smile at the absurdity of such an injunction, and the pomposity of functional theories of humour.)

Useful or not, laughter certainly adds joy and value to our lives. To regard this most human of propensities as the product or by-product of an evolutionary process controlled by chemical accidents and natural selection seems to me extraordinarily implausible. No theory of evolution can afford to ignore the Marxist aspect of man's way of life. (Not Karl Marx, of course.)

The strong tendency to ignore or discount humour, as a subject worthy of discussion, unites the diverse philosophical families of scientific positivists, linguistic analysts, existentialists, and phenomenologists. Yet humour seems as admirable an aspect of human mentality as its more solemn and sublime creations. It is part of the art of civilized living together. Though it can be hateful and spiteful, it is more naturally conducive to warmth and friendship. There is no form of human expression and communication and no art that may not be enlivened by wit and humour. Non-verbal, non-verbalizable jokes take many varied forms. Since jokes are mental achievements, they provide strong evidence that there exist forms of thought incapable of adequate linguistic formulation or translation. One may tell a joke, but one may also play a joke, or paint a joke, or build a joke, or dream one. It is an interesting question, though not relevant here, whether the different kinds of humour and styles of wit expressed through different media can be adequately cross-categorized and characterized, or whether there are medium-specific forms of humour. There exist some expressive styles, common to different media, which appear highly restrictive in the range of humour they permit or facilitate. Expressionism, for example, favours mainly the more grotesque and even neurotic forms of humour. A comprehensive account is lacking of the varieties of humorous experience and expression.

One of the greatest values of humour is the contribution it makes to man's knowledge, especially his self-knowledge. Without a talent for self-mocking, men fall easily into hubris. It is good to be able to laugh at the absurdity of human all-knowingness and claims to infallible wisdom, and the entailed absurdity of ambitions for total control over the environment, or the social and political domains. On earth only man is absurd, a fool even in wisdom, and man alone needs the corrective of self-laughter. If animals could laugh, what would they find to laugh about, apart from man? Humour helps define our common *humanitas*, our unique follies and unique remedies for folly. It is intimately related to human freedom and purposiveness, and the capacity for evaluating the experiences of life. Humour can be a means of freeing us from some of our self-illusions, and thereby facilitating wise choices. The pompous and pretentious are always with us, as permanent possibilities of each one of us. Philosophers especially need the protection of humour to save them from hubris in their imposing projects, in which they are almost certain to fall short. Socrates is a good model.

Can humour help us in any way in the understanding of the non-human world? This is very doubtful. But it may just be worth recalling that Greek Comedy as an art form probably developed from fertility celebrations which featured the wearing of masks. This 'Old Comedy' had its origin in burlesque and mime, and miming has had general ritual and humorous significance in different cultures. Children generally love it. Without this addiction to miming, it is possible that men would have been less disposed to identify instances of mimicry in other forms of life. If we are amused or intrigued by these, this seems as good a reason as any for studying them, as examples of the clash between Appearance and Reality, an incongruity which underlies much human humour, and pretending. The games of men resemble the adaptations of animals—sometimes.

Man's sense of the incongruous and the absurd may be linked, to a certain extent, to his rationality or logicality. Miming, for example, depends for its impact on the Principle of Non-Contradiction, that it is impossible for an A to be a not-A. We can laugh at the absurdity of an individual pretending to be something other than itself, only because we are logically aware of the impossibility of this. With a different logical (or illogical) system, miming might be impossible. The logical basis of mimicry is Aristotelian.

If it be true, or at least conceivable, that men's sense of humour makes it easier for them to identify and be interested in the phenomenon of organic mimicry, then in this regard humour might be viewed as a mode of perception or selective modification of perceptual interest. In any event humour, in one of its aspects, can be interpreted in this way, in the case of man's perception of man. A man utterly without humour misses a great deal that goes on in human societies. What is involved here is not the detachment of a Martian observer, but an active participation and perceptual enjoyment. The significance of humour cannot be appreciated if one refuses to be amused or is unable to respond. This must be one of the hazards facing anthropologists. Never does one feel more excluded than from laughter one does not comprehend. One of the games of children is to have secret jokes from which the uninitiated are debarred. Such things help to make the social life of children closed to adults.

Another field of interest, that may have been opened up or strengthened by the love of practical jokes and conjuring tricks, is that of sensory illusions. ('Illusion', of course, derives from the Latin word for play.) The philosopher discussing the question of the reliability of sense-perception, in the context of walking-sticks that look as if they were bent when dipped in water, a kind of inorganic mimicry, may be indebted to his earlier childish love of tricks, spoofs, and hoaxes.

But nothing that we know or surmise about the value of humour for human life and awareness can be said to throw much light on the question of its origin or unique development in man. In this respect humour is in much the same position as other valuable mental endowments of man whose contribution to human survival and reproduction remains conjectural. And the notion that man's gift of humour has been developed by an evolutionary process explainable in physico-chemical terms seems to me unintelligible, even comically so. For it is usually comic when some learned account is given of a familiar phenomenon, in which it is manifest that the learning is deployed and displayed without any real contact with the phenomenon in question. Such theories, far from explaining humour, may have to submit to its judgement. 'The origin of humour': may this not constitute another possible limit of human understanding? Perhaps all our attempts at explanation will be so many variations on the theme *docta ignorantia*. Such ignorance about the origin of humour, and other excellences of human endowment, seems far preferable to the diminution of ignorance that the hypothetical deprivation of such goods might induce. Knowledge is not the sole or overriding good for man. It is surely better to be richly endowed beings, even if self-comprehension is blocked thereby.

The difficulties that challenge scientific materialism in the explanation of biologically surplus values seem to me as serious as the traditional Problem of Evil for religious believers. These deep dilemmas, from which there is no clear escape, should make us all modest, acutely aware of the limitations of all ultimate explanations and philosophies of life.

Such awareness tends to be partial and one-sided. The problem of surplus good and suprabiological values is little stressed, and appears to lack the full canonic recognition long accorded to the Problem of Evil. Why should this be so? Are most of us, whatever our beliefs to the contrary, still under the Christian spell, so that goodness can only with the greatest difficulty be discerned as a problem? (The very phrase 'The Problem of Goodness' sounds perverse and paradoxical.) Or is it that materialists are less self-critical than theists, and less forced on the defensive by strong outside criticism? Is the neglect of this question a form of cultural blindness (problem-blindness)? Possibly. Like doctrines, certain types of question may be culturally vetoed. Philosophical questions are especially liable to changes of fashion.

But a stronger pragmatic factor in this emphasis on evil may be that life's ills make a more profound impact than its blessings. The diminution of evil is usually felt to be a more pressing problem than the maintenance of good, which may easily be taken for granted. (Some confusion between practical and theoretical issues may underlie the feeling that it is perverse to speak of good as a problem.) The great preponderance of evil over good may be inferred or assumed from the keener awareness of the former. Murder, pain, and disease compel attention, as peace and health do not. Of the latter, one might say that its *esse est non percipi*. An hour of agony may lodge more firmly in memory than years of bodily ease. Analogies to this may easily be found in the selective 'social memories' of nations, for times of tribulation.

8 Death

Among living beings, only men, it seems, know that they are going to die. This is a paradox, in view of the important life-regulating consequences of death, whether these take the form of death-resistance or death-acceptance. Reproductive behaviour and care of progeny are death-adaptations, appropriate to the impending disappearance of the parents, if the species is to survive. The link between death and reproduction is sometimes dramatically pointed, as when, it is said, some female arachnids devour their partners after copulation.[1]

How, without knowledge of death, is death-adapted behaviour possible? How are posthumous concerns possible, as when short-lived insects lay their eggs near a food supply for the benefit of the young they will never see? How can the negative fact, unknown to them, of their future non-existence, influence the death-appropriate behaviour of insects, and other groups? No safe answers are possible, only guesses. It is 'as though' their deaths had been foreseen, though not by them; rather as certain prisoners during the war were forced to dig their own graves without realizing they were about to be shot and shovelled into the trenches they had completed. They were surprised by death, despite their death-oriented behaviour. The latter, however, unlike the behaviour of insects, was not instinctive. We do not know how posthumous instincts, concerned with posterity and the future of the race, originated and developed. It is rather superficial to think the general problem of death-adaptation, of ways of safeguarding the survival of the race despite individual deaths, can in principle be resolved by reference to genetic changes and their selection. For genetic systems, by virtue of their role in reproduction, are themselves, *ex officio*, death-adapted. The problem therefore is a very fundamental one, pre-genetic in its implications. Death-adapted replication is as

[1] *Post coitum omne animal triste.*

old as life. Life would not exist, as we know it, were death
not inevitable, in the 'long' run (and avoidable in the short
run).[1] Especially in its more primitive forms, its central concerns
are death-resistance and reproduction. Metaphorically, one could
say that death constitutes part of the collective unconscious of
living beings, informing their activities in a wholly unconscious
way. It is not surprising that men, having discovered this great
life-principle and secret, attempt to forget it. It is knowledge that
is difficult to live with, if only because of the tension it sets up
between the urge to resist death, and the urge to acquiesce in one's
death by the raising of one's successors. No such difficulties exist
for animals, as they are unaware of any tension between the
aims of self-preservation and self-reproduction; between resist-
ing the inevitable and conceding it.

Once again it is 'as though' the applied results of the foreknowledge of death had been embodied in the instinctive
behaviour of animals—without the knowledge, or the possibility
of acquiring it. In any case, their dying is not an inert terminal
event, like the final flicker of an expiring candle; it is inadequate
to say that organisms are mortal, or to regard their mortality as
a passive happening or process of wearing out, as a carpet or
machine may wear out. Their mortality is, as it were, active or
alive, pre-intimated or reflected in their behaviour, a pervasive
life-ingredient. In this sense, they are not simply mortal, they
'mortalize', being active with respect to their ends and posterity.
Their ends are (implicit) in their behavings.[2]

Even if animals could have acquired empirically the knowledge
that they were going to die, it is impossible that their death-
adapted behaviour should have been dependent on this knowledge.
Their line would have become extinct before the knowledge
could be obtained and applied. This knowledge, however,
makes possible new forms of death-adapted behaviour, unique to
man. The recognition of this necessity confers a certain freedom in
the utilization of the knowledge, and ways of 'coming to terms'
with it. Nowhere is the line between animals and men more
sharply drawn than here. This knowledge plays a decisive role

---

[1] The theoretical immortality of micro-organisms that multiply by fission is
Utopian, outside the laboratory. It presupposes an ideal environment, organic
and inorganic, in perpetuity.

[2] This does not apply to individuals that play no part in reproductive processes.

in human history and historical awareness. Animals have no conscious history, not realizing how they are related to past and future generations. They are unaware of themselves as individuals having a beginning and end. In this respect they are more like inanimate objects, or the sun and moon, than like man. Man lives among animals like a grave-digger among coffins, that are death-adapted and death-ignorant. Ignorance concerning death is also ignorance concerning time, and an individual's life-time or span. The difference between a butterfly-span and a 'planetary' span is nothing for a being incapable of span-consciousness. It is difficult to think about the age of a farm or household animal, or about the age of the universe, without projecting human concerns and anxieties. Will it die soon? Is it running down? Ageing, as self-concern, is peculiar to men. These self-centred anxieties spread out in wider circles and concerns. Even our artefacts may be assigned a death date. To be aware of one's finitude is potentially to be aware of the passing of everything, to survey the universe *sub specie mortis*.

Among finite beings, only one who knows that his visit is a temporary one can form the conception of some greater continuum or world order, a more durable framework than the limits of individual life. No animal has conceived the time before his birth and after his death; no general concept of time has been formed. To realize the limits of individual existence is to think beyond them, an expansion of mental horizons that is unprecedented on earth before man appeared. This awareness seems to be related to the possibility of developing a religion and a cosmology, in which the limits of individual existence are transcended. If so, the thought of death may be said to have played a constructive role in the development of human culture and human art.

The law of death is one of those apparently rare laws that seem to admit no exception whatsoever, no statistical loophole. Otherwise the uncertainty would be unbearable. Knowledge of this law must have come to man through his language, which is, in a sense, more durable and 'wiser' than any individual user of it. Through it one is able to outreach the experiences and memories of one life, and learn ancestrally. If one were confined to the experience of one life, one might opine that all men die, but conclusive evidence would be lacking; one would be dead first.

Knowledge of death as universal is the first great fruit of historical knowledge, mediated by language. This is possible because a language, and the information conveyed by it, have a far longer life or history than that of any individual or generation.

Human societies depend on the foreknowledge of death and would hardly be conceivable without it. Institutions that are not death-adapted, that make no provision for their continuous manning in the face of death, disintegrate and go under like a boat that has lost its crew. (Temporarily the empty throne and vacant office may remain functional socially, if it is believed that they are manned or about to be—rather like the influence emanating from the empty thrones of some mythical pantheon.) The young are trained to 'take over' when the time comes. Socially necessary 'functions' are discharged by successive 'functionaries'. There are no permanencies; the one principle of permanence is that all positions are occupied temporarily. Without rules of succession, chaos results. 'The King is dead: long live the King.' Even in societies based on the hereditary principle, this is not automatic or biologically necessary (like the functional differentiation in an insect society), but constitutes one possible solution to the problem of maintaining the continuity of a society and its institutions, in the knowledge that death is a permanent possibility, and an ultimate certainty, for all.

The most significant moral, political, and legal issues, and the most dramatic forms of art, spring directly from the foreknowledge of death, and its many possible uses and abuses. Without it, I do not see how any animal or group could be thought capable of suicide, or genocide. The knowledge of death makes possible new ways of dying, and living. We cannot choose not to die, but may choose, any day, to live no longer, if we have the nerve, and our resolve is not self-defeating, by mobilizing crushing opposition. This freedom has obvious social implications, but is embarrassing to legislators, as no potential self-killer is going to be intimidated by threat of punishment, of the ordinary (unmetaphysical) kind. Divine sanctions have to be invoked—if any. The harshness of these may be related to their total unverifiability (at the time when the act is deliberated). It is assumed that suicide cannot be repeated, to escape from hell. (The only system that allows repeated suicides—the system of Reincarnation—insists that no suicide is final, preventing rebirth.)

Foreknowledge of death has other, less metaphysical, consequences for human life and behaviour. By throwing open to consciousness the time after one's death, it makes possible the desire to be represented and remembered posthumously, to leave behind some lasting legacy. Consciousness of death is the condition for seeking posthumous ends, even though the presumed death of consciousness for the individual makes it impossible for him to know if his aims are realized, or to bask in their enjoyment. Such posthumous behaviour may appear either as the absurdest vanity or as the purest altruism—the act of an old man planting trees, secretly. Whatever the motivation, future generations are on the whole the gainers from this unique human concern for making something that will outlast a lifetime. In this sense, human culture is consciously death-out-reaching and differs radically from the posthumous instincts of animals. Animals have no cities, only coral reefs and the like. Having no sense of a lifetime, they cannot realize that this limits and encloses an individual's acts, without precluding their possible posthumous significance. (The posthumous fame of some animals does not run among animals.)

Only men can impose on themselves goals and tasks whose realization they attempt to time, however roughly. Foreknowledge of death is the main source of the sense of the value of time, as something limited, in relation to the possible acts and projects of each individual. The notion of work, or the performance of finite tasks, 'weighed up' and 'measured', is bound up with the idea of limited time. Only men can be said to 'fritter their time away'. Not knowing the day of one's death (but knowing one has not got 'all the time in the world'), time-gambles are unavoidable: man proposes, death disposes. Mozart, with his daily premonitions of the closeness of death, completed his last three symphonies in eight weeks and worked feverishly on his unfinished Requiem. Wagner, one may be sure, had no such urgent premonitions when working on *The Ring*, but must have assumed that his life-span would allow its completion, through great interruptions and developments of the original plan. He staked everything on this and won. Otherwise he would be mainly famous for his fragments, for his aspirations rather than achievements (like the heroes of his dramas). Less daring artists reduce the risks by becoming choppers, cutting off single works from those that are to follow, even though a single ruling idea or theme may inform these. Mallarmé, despite his preoccupation with his '*Grand Oeuvre*', '*Le Livre*', published single poems of uncertain relation to the latter. Master of mystery in his achieved poetry, he deepened the mystery of his poetic purpose by leaving life, and '*Le Livre*', prematurely. *Tel qu'en Lui-même enfin l'éternité le change.*

The abolition of death and the consciousness of death would mean the devaluation of time, its conversion into an indifferent medium or equable unending flow.

If the choice of finite tasks and pursuit of definite goals with approximate deadlines is bound up with the notion of death and a limited future, a life without end would presumably be a life without definite ends, an endless drift through unhorizoned aeons. If human purposiveness and its associated values are thus connected and articulated, the impression that these values and purposes are, *in principle*, nullified by the inevitability of death, is without foundation, since they are death-dependent. All human activities—even the most 'creative'—are framed and clarified within this imposed horizon, even when they look beyond it. That is the general principle. In practice, a person's life, in terms of his purposes and projects and prospects, may be wrecked prematurely, by an 'untimely' death. But it may be wrecked by many other factors. Death is not the unique destroyer. Without it, we should not know how to live or act. We would be totally out of our element, adrift on a boundless sea, disconcerted by an infinity to which we are not attuned. Death saves us from a fate worse than death. Horror at death is unilluminating about alternatives. Is there about it something of induced hysteria or of a self-indulgent sentimentalism based on a distorted view of human possibilities? These appear to be death-adapted, appropriate to a finite term of life and span of development, and to the finality of death.

One of the strongest feelings opposed to the finality of death is grief. If grief has the last word and the hope of reunion is vain, this seems a cruelty that destroys the credibility of a benevolent God. The grief that may turn a person Godwards should turn him away for ever, if there is no surviving of death.

This is an intelligible raging against death, but the sense of moral outrage is ambiguous. Grief is the instinctive reaction to the loss of someone loved. If the loss is only temporary, the sufferings of grief are unnecessary, gratuitous, and absurd. Grief is appropriate or 'justified' only if there is no possibility of a reunion. Otherwise this instinctive reaction and anguish at the disappearance of a person would be deceptive and delusive.[1] A God that permitted such unnecessary and uncalled-for suffering would be a cruel deceiver. He would be no better than the Duke in the play *Measure for Measure*, who falsely makes Isabella grieve at the death of her brother Claudio, who is not dead. 'Unhappy Claudio! wretched Isabel! Injurious world.'

Injurious world. On the assumption of the finality of death, grief fits into this world as the inevitable consequence or concomitant of love and death. If there were no grief for the dead, there would be no love for the living. If there were no love, there would not be so much to live for. And if there were no death, there would be no love, at least as sexually expressed. If there were no death, life would be inconceivably different, and intolerable for people like us. These all hang together, the best and worst features of our existence organically linked. It is futile, even in imagination, to wish away the worst and hold on to the best. The blend is indissoluble.

If the dead stay dead, what value or significance can be attached to fidelity to the dead? One cannot be faithful to a corpse. Are all promises, vows, and obligations dissolved with the final dissolution of the promisee? How can the dead be thought to have any rights or claims on the living? They are beyond human protection, and the need for protection. They cannot be injured, cannot be defended from injury. How can the principle stand, that it is wrong to defame the dead and misrepresent their lives and actions? Those dry bones will not be hurt by defamation. They are as unslanderable as the weather. Is not this respect for the dead a senseless superstition, unless the dead are believed to be mentally alive and sensitive to injustices inflicted on them after their physical dissolution? With growing scepticism about the possibility of surviving death in any form, should these

---

[1] Ben Jonson's couplet: 'He that feares death, or mournes it, in the just,
                         Shewes of the resurrection little trust'
may be capped: ''Tis in our nature to shew little trust.
                       If nature lies, then how can this be just?'
This is too pert, but crystallizes the issue.

ancient superstitions be abandoned? Or should one think, on the
contrary, that the universal validity of the moral principle
involved testifies to some form of life after death, as the necessary
presupposition of the principle?—an 'ought' that implies an 'is'?
May not even those who flout this principle, and pursue their
enmity beyond the grave—by mutilation of the corpse, or
defamation—be testifying to their belief that the spirits of the
dead can be tormented? Was it necessary that the ghost of
Hamlet's father should appear (or appear to appear) to Hamlet,
uncertain of the reality of survival after death, to show him that
he still had filial duties? If so, the ghost scenes in *Hamlet* would
have a moral point lacking in other ghost scenes, such as the
statue scene in *Don Giovanni*, where the slain Commendatore
takes over his own revenge.

Actually there does not seem to be much difference in principle
between obligations to the dead, on the assumption that there is
no after-life, and obligations to the living. If the unconsciousness
of the dead dissolves all obligations to them among the living, and
not merely *some* obligations, on the ground that the dead are
unaware and unworried by broken pledges and other betrayals,
then for similar reasons a pledge made to a man who disappears
into an enclosed monastic order for life, cut off from outside
communication, is morally annulled. If a once famous person
becomes incurably insane and understands nothing of the slander-
ing calumnies made against his former life, this does not excuse
deliberate defamation. 'On the contrary', one may think. The
deafness of the accused cannot excuse the insolence of his accusers.
A wrong done against a dead man is a wrong against him as he
was and lived, not against his mouldering corpse. It does not
become right or morally indifferent merely because he can
no longer know of it. His death, in this sense, is morally
irrelevant.

There is, naturally, a difference in range or content between
obligations to the dead and the living. Some obligations, for
example to the sick and oppressed, can only be owed to the
living and are dissolved at death. Others, that can be owed to the
quick or the dead, are death-indifferent. On the assumption of the
finality of death, I do not think there are any unique obligations
owed only to the dead and beginning after death, for which no
analogue exists in the case of the living. This would seem to

follow from the retrospective nature of such obligations, which are continuations or extensions beyond death of obligations that were owed, or might in certain circumstances have been owed to a person before he died (such as gratitude).

Some alleged obligations to the dead presuppose that they have survived their deaths, and are unique to them: such as the obligation to pray for their souls or to perform certain last rites that will ease their departure for the next world. These are prospective services, valid only if the dead have a future before them which might be affected by ritual acts.

A different sort of prospective obligation is to the unborn. Such obligations differ in content from retrospective obligations to the dead, but have the same general property of being obligations on the living to persons who are not living at the relevant time. People may be wronged after their deaths, or before their births even. Morality jumps across generations, in that uniquely human bond of association that relates the living, dead, and unborn.

Fidelity to the dead, unless they survive their death, may still seem to be based on a kind of illusion, of the felt or imagined presence of the dead one: the result of an inability to adapt to the finality of his disappearance, to the fact that this tie has been finally severed. One could speak of phantom ties and phantom obligations, by analogy with the phantom limb whose presence continues to be felt after it has been cut off and died. This may be so in some cases, but it would not explain the sense of obligation to 'do justice' or 'deal fairly' with the actions of some remote historical figure, with whom no personal ties have been formed. This might be regarded as a secular version of the Last Judgement, with historians taking over one of the functions of God. However this may be, and however strange and enigmatic the behaviour of the living towards the dead may appear, it is not possible to draw any inferences, one way or the other, from this behaviour concerning the possibility of a life after death.

This possibility cannot be finally disproved, however indispensable may seem the integrity of the body to all aspects of personal existence. With what right could anyone have dared believe, before the event, that a deaf man would be capable of creating music surpassing that of all but the greatest composers with perfect hearing? Beethoven himself appears never to have lost faith in the continuance of his creative power, but many of his contemporaries, misled by a very reasonable prejudice, were unable to accept his later music at its true value, ascribing their own inner deafness to his outer deprivation. Beethoven of course was not *born* deaf. But some of his most venturesome and exploratory sound-weaving came after his deafness, such as the Grosse Fuge of opus 130. Such *a priori* inventiveness represents a barely credible attenuation of sensuous requirements, a potent refutation of any empiricist concept of invention or imagination! Musical invention is not the product of what is heard, but the producer. Those that lack ears to hear, let them invent! We are not least indebted to Beethoven for his classic demonstration of the inadequacy of empiricism.

It does not of course follow that because a man's creative and imaginative talents can survive the loss of his senses, unimpaired or even strengthened by the loss, they could equally survive the total loss of his body! But an absolute denial that this is possible seems unwarranted, however improbable it may now seem.

Beethoven's ability to create marvellous music that he could never hope to check with his outer ears seems more miraculous to me than many of the alleged para-normal phenomena studied in Psychical Research.

Prevalent disbelief in a life after death may not rest primarily or exclusively on the absence of solid evidence for survival and the suspect nature of alleged messages from the dead. There is a similar lack of solid direct evidence of living and rational beings on other planets. Yet many scientifically minded people are disposed to take seriously this latter possibility—or even 'probability'. Belief and disbelief in these two cases seem to depend on certain general ideas about what is likely or 'reasonable' and what is not. In one case general physical considerations are taken to close certain possibilities of existence, and in the other case to open them. Physical reasons make it seem unlikely that mind or spirit could continue without a body, or with the old body reconstituted, or with a brand new body (reincarnated). In the second case, physical theories about the nature and origin of life on earth make it possible or probable that on some other planets where physical conditions are favourable, life will spontaneously originate and evolve. Against this general theoretical background, it is intelligible that the absence of solid direct evidence in both cases should carry different weights.

Let us now assume that death is the final end; does this imply that one's life is made null and void, by its final cancellation? How can the closing of a life be believed to nullify, retroactively, the antecedent experience and attainment? How can the destruction be total? (The same nihilistic paradox underlines Prospero's premonition that the globe itself and all which it inherit, will be dissolved, as though it had never been.)

Surely all that is 'destroyed' at death, that is of potential value, is the possibility of further individual growth and attainment, by whatever criteria these may be judged; and except in the case of grossly premature deaths, such possibilities may be slight, and are never unlimited? Destructiveness is a relative conception, that should not be predicated of death absolutely, without qualification and discrimination. Lamentable though premature death be, its incidence and constant possibility are bound up with the general mortality of individual organisms; without miraculous intervention, it could only be abolished through the abolition of death as such. And this would presumably be catastrophic, for beings constituted as we are. The fundamental terms and conditions of life are not arbitrarily imposed on us as individuals, but are immanent to our whole being.

Birth and death define our nature as finite individuals. Death is regarded as bad and birth as good, generally (except by total pessimists), but both are limiting brackets of individual existence. Although the necessity of being born is as firm an indication of one's temporariness as the necessity of dying, it gives rise to no great concern or despair; on the contrary, men celebrate the anniversaries of their birth, and do not repress all thought of this event, as they repress the thought of their demise. Exceptions to this usually derive from the connection between being born and having to die in the not too remote future, the premonition given by signs of ageing. Apart from this connection, there is little fear that our lives are made meaningless by having to start at a particular time. In this direction we are very well attuned to our finitude, and undisturbed by the retrospect of non-existence and prenatal nothingness. To be so disturbed, to an equal or greater extent than at the prospect of future extinction, would generally be diagnosed as an extreme phobic eccentricity and aberration, the mark of a nihilistic invert, for whom time was running out backwards, and fears reversed themselves; fears which no promise of immortality could extinguish, since this could never remedy the aching void of prior non-existence. Only a doctrine of 'innatality' or prenatality could remedy this, or belief in an infinite sequence of rebirths, with no first term.

A religion which underwrote belief in the individual's prenatal existence, but denied the possibility of future survival after death, would fail to satisfy normal human aspirations and would 'have no future'. Buddhism and Hinduism exceptionally offer both. But what of the ultimate extinction of Nirvana? This is only the end of separate individual existence and suffering, and cannot be identified with total annihilation. In a way, the concept of Nirvana brings Buddhism in line with most other religions, in setting up a future state as the ultimate destination of the soul. Temporal asymmetry is transcendentally restored, and the privileged significance of the future reasserted. In its restless migrations from body to body, there is a goal, however remote, for the soul, a final overcoming of the past. (How far this doctrine, or that of reincarnation, is consistent with the Buddhist denial of any continuing soul or self, must be left aside here.)

A similar sort of Futurism is reflected in men's attitudes to the survival of organizations, traditions, forms of expression and belief, in which they are involved as individuals. They are generally more disturbed by the idea of the future disappearance and extinction of their tribe, nation, race, religion, language, art, than by the thought that these have not always existed. The notion of the universe running down may be found depressing, unaccompanied by any concern at its finite age from origin until now.

Language is impregnated with this temporal value-prejudice. Contrast the elegiac tone of such terms as 'transient', 'fleeting', 'evanescent', 'fugitive', with the celebratory note of 'nascent', 'burgeoning', 'new-rising'. These value-opposed sets of terms equally characterize the temporariness of existents, viewed from different temporal directions. 'To be', in this world, means 'to be temporary', except in the context of concepts and truths believed to be timeless.

Since 'temporary' has acquired its own pathos, largely through its half-meaning of 'terminating' or 'terminable', it would not too readily be used of stars and stones, where pathos may seem out of place; still less of symphonies. These are temporally consummated developments of certain musical ideas, and escape the pathos of the temporally unfulfilled, the 'cut short'. The relevant sense of 'finite' here is fully realized or finished. Hence the appellation of Schubert's 8th Symphony. It may be difficult to decide, in the case of a work like Schoenberg's *Moses and Aaron*, whether it is fully realized, despite its appearance of incompleteness. *Mutatis mutandis*, similar difficulties may be experienced in the case of individual human lives, to an even greater degree. This analogy may be rejected, but I think it deserves fuller consideration than it has received.

To return to the value-prejudice which rates future existence and non-existence as far more significant and concern-worthy than prenatal existence and non-existence: why should there be this difference of valuation? Why should forward finitude seem bad, when past finitude is tolerated or regarded with indifference? Why react so differently to these two indissolubly connected aspects of temporariness? Are we the fools of time, if we suffer (the word is apt) from the delusion that a future limit is essentially different from a past limit, and justifies this subjective conflict of response? Or does this very general, if not universal, difference of valuation have an innate origin? If so, could it be maintained that we would not be predisposed or pre-instructed to value future survival far more highly than pre-existence, if this innately grounded preference were not in accord with the arrangements of providence for our future destinies, arrangements themselves temporally biased? Is God a futurist, and are we made in His image? Or is it perhaps the other way round, with the God-makers foisting on to God their own futuristic prejudices?

The suggestion of an innate origin for our overriding pre-occupation with future survival seems plausible, though not these other speculations. The desire for life everlasting may be a manifestation of the life-prolonging instinct. This would make intelligible the different valuation set on future survival and pre-existence. A life-prolonging instinct is necessarily forward-looking, since the past is beyond its range of possible influence. In their expectation of another life, men have transcendentalized this instinct, and taken over its futuristic tendency (which it shares with all other active dispositions and interests). We have no goals for the past.[1]

A philosophy of death ought not to be the slave of instinct, ought somehow to be capable of detaching itself from this deeply ingrained practical orientation and of regarding past and future non-existence similarly. If the awareness of my backward limit does not drain my life of all sense of its value, why should awareness of my coming death be allowed to do so? Existentialism is especially liable to overvalue future existence, perhaps because of its strong activistic bent and characteristic emphasis on the individual as his own 'project' drawn to his future possibilities.[2]

If my life had no backward limit, I should doubtless have had enough by now. If it has no future limit, escape is impossible.

In some ways all this concern with death may appear over-specialized, narrow-minded, and parochial. If one leaves out of account for a moment all one's personal and human fears and feelings, and even all one's organic empathies, one may see that the death-theme, in certain of its aspects, forms part of a greater and indeed universal issue: the nature of time, and in particular what is involved in the passing away of all that is temporal. The childish question 'What's become of the past, of all that's passed away?' isn't easy to shrug or sophisticate away. Has the substance of the remote past, in the fullness of its being, evaporated to nothing, or to the few faint tracks and traces that may still be observed? Historians assuredly are not studying nothing, or the mere bones of the past, even if they start from these. They are studying selected aspects of the full temporal and historical reality. Apart from mathematics and theology, there seems nothing else to study.

The historically assumed reality of the past may be interpreted to mean that past events, including the life courses of the dead, may

[1] The concept of another life as morally connected with this one, that one is judged and gets one's deserts, strongly favours the futuristic concept of im-mortality (or post-mortality). Even in the doctrine of reincarnation, the emphasis lies in improving one's lot in future incarnations by one's present conduct.

[2] A not dissimilar view of life is that of the climber, whose life-plan is ruled by a barely visible series of peaks to be conquered, each higher than the last but with no final consummation or Everest. This produces a sense of strain towards the future and dissatisfaction with the past, which represents a lower level of aspiration. In its incompletability, this life-plan has something in common with the labours of Sisyphus. A facile generalization would be that Western man lives in a Sisyphean society.

have passed from the field of view and daily scene of later human (or animal) existents, and from their range of possible influence and intervention; but have not passed out of being, out of the trans-temporal field of relationships in which the present grows out of the past and depends on it even in its power of deviating from it. The central conception is that of the interdependence of past, present, and future, so that one could not cast doubt on the reality of any of these independently of the others. (This might be regarded as the 'ontological' formulation of the corresponding epistemic thesis, that memory, perception, and anticipation are so interrelated logically that scepticism about one would imply scepticism about the others.) The reality of the past, on this view, is fully compatible with its 'latency', with its being hidden from the field of view (and of possible action) of later existents. As spatial distance between observer and object may prevent observation, so may temporal separation. Why assume that the past has passed out of being, simply because it is no longer observable? Observation may itself be dependent on memory, which normally carries conviction of the reality of the remembered past. In any case, if one accepts the standard account of perception, some past states or events are observable, their observability in principle being a function of the spatial distance between observer and observed and the speed of light.[1] How could we see past aspects of the universe, if these have ceased to be? Are we seeing ghosts? This raises problems of perception too complex for discussion here.

Like everything to do with time, this concept of the reality of the past, whether observable or not, is extremely difficult to formulate and comprehend, and may be wrong or incoherent. But unless one identifies reality (or realities) with ever-changing slices of a cut-out present moment, one needs some concept of the reality of the past that will hold things together and save the universe (conceptually) from a perpetual perishing. How this concept of the reality of the past should be interpreted seems to me one of the most important issues for any world-view that 'takes time seriously'. The idealist notion that the past is nothing apart

[1] This principle has its temporal obverse, in the case of man's possible action at a distance. The more spatially remote the region to be acted on, the longer it will take for the change to be transmitted, for any given form of human technical intervention. We perceive the past, we change the future. We may be dead before our actions reach their target.

from human consciousness of it (which seems to be implicit in the significance attached to the raising of monuments and memorials) has no more to commend it than the idealist position in other domains.

Time, in one of its cosmological 'functions', might be compared to a segregating device that prevents waves of existents from jamming one another in spatially restricted regions. Temporal separation is the functional complement, or supplement, of spatial separation. Temporal separation of existents prevents mutual interference and incompatibility, as within limits the spatial distancing of contemporaries does. We ourselves use time in this way, when we run out of space. In a small room, we hang our pictures serially, when they cannot all be shown at once, or when they are mutually incompatible.

Dare one think of the passing of time as separative, rather than annihilative, of all temporal and temporary existents?

9 Coda: Atomism and Atheism

Modern theories of life and its evolution embody some of the same basic ingredients as one finds in the ancient Atomists, such as Leucippus, Democritus, and Lucretius. Recent developments have narrowed the distance between ancient and modern themes. The title of Jacques Monod's book *Le hasard et la nécessité*, is taken from Democritus: 'Everything existing in the Universe is the fruit of chance and necessity.' An unfortunate title, possibly, in its implied reification of these abstract notions and attribution of productive power, an implication Monod does not wholly avoid.

It would be an exaggeration, but a pardonable one, to say that no leading principle of significance separates Monod from Lucretius, that the former merely knows more chemistry. Philosophically they are close kin. From their Atomism[1] both deduce the necessity of Atheism (with idiosyncratic qualifications in the case of Lucretius). Lucretius would certainly and with great delight have hexametered Monod, Crick, and Watson, had they been ancient Greeks or Romans. The double helix of DNA would not have defeated his inventive Latinity. Lucretius was the first great poet of scientific materialism, in fact the only one so far. I believe his poetry is superior to his philosophy, and certainly more original. As Eliot pointed out, poets do not usually invent philosophies. Lucretius certainly did not (nor Eliot himself). This is not surprising, as genuine philosophical originality is very rare indeed, much rarer than is commonly supposed, and the chances of it going along with poetic ability are slight, if not unknown.

[1] Monod thinks any dissatisfaction with exclusive reliance on the analytical methods of molecular biology is stupid: in a mere twenty years these methods have dissolved most of the 'mysteries' of the older Biology (*Chance and Necessity*, pp. 37 and 80).

Whatever the Greek and Epicurean sources tapped by Lucretius, anticipations abound of evolutionary and genetical ideas. I shall quote a few:

The passage of time changes the character of the entire universe, and one state must give way to another. Nothing remains the same. Terrestrial states succeed one another through time. What the earth once bore, she can bear no more, and what she could not bear, she now bears. Many are the species of animals that must have perished, unable by reproduction to forge a posterity. For whatever you see feeding on the breath of life, must have been protected and preserved in existence, from the earliest time of origin, by its cunning or courage or speed of movement. Animals that are not endowed with such gifts, so that they could neither live freely from their own resources, nor come under human care and protection by reason of their usefulness to man, became the profitable prey of other species, fettered by their own fatal weakness, until nature brought the species to extinction.[1]

Hereditary patterns remain stable to such an extent that richly coloured birds manifest in the succession of generations the bodily markings that are peculiar to their kind. To account for this, they must carry in their bodies an invariant physical substance.[2]

Of course, mingled with such striking anticipations are other ideas that are clearly inadequate and wrong. But Lucretius held fast to the first principle of a thoroughgoing molecular biology, that everything in the structure, behaviour, and experience of living beings, including man, could in principle be explained as the result of the activity of their component parts or atoms in their mutual interactions, and that the particular atomic constellations that constitute individuals of different kinds originated by chance, and their species survive or die out according to their hereditary endowments, their fitness, and reproductive ability.

Lucretius was certainly dazzled by Atomism, as by an *idée fixe*, and saw everything atomically (even the Gods). But unlike the theoretical and speculative Greek Atomists, he valued Atomism above all as a bridge to Atheism (in all but a nominal sense).

[1] *De Rerum Natura*, Bk. 5, ll. 828–77 (excerpts).
[2] Idem, Bk. 1, ll. 588–92.

Atomism made unnecessary the hypothesis of a Demiurge (atoms being sempiternal presumably). Lucretius hated religion for the frequent cruelties of religious practice, and for its belief in an after-life, dread of which held living men in torment and drove away all chance of happiness during an overshadowed existence. Atomism then was a liberating science, a gay science that relieved men of the miseries of religion. The fervour of this belief turns *De Rerum Natura* into something greater than a poeticized treatise on atoms. An alternative title would have been *De Deorum Timore*.

This missionary zeal may help to explain the astonishing confidence with which Lucretius offers atomic explanations of absolutely everything, including the Trojan Wars. He must have felt that a single exception would have opened up a crack through which the great enemy of human serenity might infiltrate. He constructs a world that is wholly self-sealed and self-sufficient, in which men can live their brief span, if not securely, at least free from fear that the worst is yet to come.

The soul is composed of the finest atoms, and like water contained in a jug, is wholly dissipated when the urn of the body is fractured. Life's end is fear's end:

> *Scire licet nobis nil esse in morte timendum*
> *nec miserum fieri qui non est posse neque hilum*
> *differre anne ullo fuerit iam tempore natus*
> *mortalem vitam mors cum inmortalis ademit.*[1]

'We can know for sure that there is nothing to be feared in being dead; that one who is not cannot be made to suffer; and that it is a matter of supreme indifference whether one might ever have been born at all, when death that never dies has seized one's dying life.'

It may seem odd that Lucretius should have troubled to compose an epic on atoms to relieve Romans of their death-fears. Much simpler ways of doing this can be imagined, hardly less effective or more ineffective. But that would have made a different poem, and a less imaginative one. The atomic vision of the universe, and of man's 'dying life' as an integral part of it, subject like everything else to decay and supersession, swept

[1] *De Rerum Natura*, Bk. 3, ll. 866–9.

Lucretius away and gave apparently cast-iron backing to his irreligious convictions. He needed a complete Credo.

Theoretically, an all-explaining Atomism is as objectionable as an all-explaining Theism or any other Ism. Belief in Total Explanations places an absurd over-confidence in the powers of human thinking and wise selection of 'authorities', in the case of derived dogmas. Einstein once said that the genuine scientist 'attains that humble attitude of mind towards the grandeur of reason incarnate in existence, which in its profoundest depths is inaccessible to man'. 'The finest thing we can experience is Mystery. It is the fundamental emotion that is at the roots of true science. Those who do not know it, those who cannot admire, those who are no longer capable of experiencing a sense of wonder, might as well be dead.'

The eternal inadequacy of human understanding rules out the possibility of total explanations. Even if one distant day all our questions in some domain were to receive thoroughly convincing answers, there might well be other questions we had not thought of or had been unable to formulate clearly. Answers are shaped by questions, but questioning itself cannot be pre-determined. Questing and questioning are the creative spur that keeps our minds moving, our theories open, and our life-goals changing. Their future course, like that of other creative acts, is not predictable.

Unpredictability has an ambivalent quality for men. It makes life interesting, exciting, and radically insecure. Monod suggests that the need for complete explanations may be innate, and that their absence causes deep inner anxiety. Though he speaks of this in the context of animistic myths, it is conceivable that his diagnosis applies to his own striving for a total explanation of 'creative evolution'. 'Chance *alone* is at the source of every innovation, of all creation in the biosphere. Pure chance, absolutely free but blind, at the very root of the stupendous edifice of evolution: this central concept of modern biology is no longer one among other possible or even conceivable hypotheses. It is today the *sole* conceivable hypothesis, the only one compatible with observed and tested fact.'[1]

Such claims to omniscience, over such a vast range and variety of phenomena, seem to me absurd. There is also a strong hint of

[1] *Chance and Necessity*, p. 110.

'animism' in the personification of chance as 'absolutely free but blind', the one creative 'source' of everything. Though Monod goes on to analyse different meanings of 'chance', the rhetoric of this passage defies analysis. 'Rigour now is gone to bed.' It is the statement of a Credo; a Monodology.[1]

So far as there is any basis for the more intelligible parts of this pronouncement, it is drawn from experimental research in molecular genetics, which shows the random nature of the different kinds of mutation so far encountered. This does not ustify the dogmatic assertion that these constitute the only possible source of genetic variation. Even the totally confident belief that all inherited characters are transmitted by DNA may need qualification in the future (see later discussion).

There seems little reason to doubt that Monod's certainties, like those of Lucretius, are partly motivated by his hostility to religion, though presumably he has different reasons for this, as fear of a life after death does not rank high among the anxieties of men today. For many people religion has become irrelevant, and science is more likely to inspire apprehension, not only for its practical but also its philosophical implications, of an aimless life set down in a desert of meaninglessness. For a poetic realization of this, one has to turn from Lucretius to Mallarmé, who in his trangest poem writes: '*Un Coup de Dés jamais n'abolira le hasard....* *Toute pensée émet un Coup de Dés.*' Exegesis is risky, but I take his meaning, or a possible meaning, to be that in a dice-thrown universe, man himself will never overcome a final aimlessness by the power of his intelligence and foresight—since thinking itself resembles a dice throw. Possibly this might be described as an example of 'retrojection'. Men often project their ways of feeling, etc., on to the outer world, but they also retroject from the latter. This may lead to dehumanization and depersonalization, and a loss of the sense of human control and responsibility. One could give many examples of this today. Our world-views do not leave us unscathed.

---

[1] Creativity has not yet been made intelligible, either in the Biosphere or the Noosphere. In default, refuge is taken in aleatory 'explanations', in both cases. In the Noosphere, a consequence is the generally disastrous reliance on aleatory procedures in music and the other arts. This suggests possible confusion between the unpredictability of creative improvisation and that of dicing, or between 'quantum' jumps and imaginative ones. Creative ideas are unpredictable, but are not the fruit of an aimed-at unpredictability.

Toute pensée vient d'un coup de dés. If man's thinking powers derive at source from random molecular events, how is it possible to place any confidence in their cognitive claims and aims? If, as Monod says, 'pure chance, absolutely blind' is at the evolutionary root of our mental equipment, how is it conceivable that true insight about our evolutionary past is achievable? By what process of 'angelization' could men have become cognizant of their random origins and spectators of all time and existence, as though from some superior and independent vantage-point? Do the Neo-Darwinians, like many other system-builders, desert the system of which they are the authors, claiming special cognitive privileges that cannot be justified within the system? Or alternatively, do they commit cognitive suicide? If our mental equipment has been formed and developed by non-rational processes, what possible grounds have we for trusting it when it infers, for example, that the chance hypothesis is the sole conceivable hypothesis? Reason totters and everything becomes logically permissible? Mere anarchy is loosed upon the word?

It is possible that Monod may have had some such questions in mind when he wrote as follows:[1]

If we are correct in considering that thought is based on an underlying process of subjective simulation, we must assume that the high development of this faculty in man is the outcome of an evolution during which natural selection tested the efficacy of the process, its survival value. The very practical terms of this testing have been the success of the concrete action counselled and prepared for by imaginary experimentation. Hence it was on account of its capacity for adequate representation and for accurate foresight *confirmed by concrete experience* that the power of simulation lodged in our early ancestors' central nervous system was propelled to the level reached by *Homo Sapiens*. The subjective simulator could not afford to make any mistakes when organising a panther hunt with the weapons available to Australanthropus, Pithecanthropus or even *Homo Sapiens* of Cro-Magnon times. That is why the innate logical instrument we have inherited from our forbears is so reliable and enables us to 'comprehend' events in the world around us, that is, to describe them in symbolic language and

----

[1] *Chance and Necessity*, p. 147.

to foresee their course, provided the simulator is fed with the necessary elements of information.

Some large jumps are made here; especially the jump from panther-hunting to scientific understanding. In a passage immediately following the one I have quoted, Monod attempts to explain the applicability to nature of mathematical notions created independently of sense-perception, by reference to the development by natural selection of the requisite mental capacities in our ancestors. Can our trust in the fitness of our mental equipment to comprehend nature by means of concepts that 'owe nothing to experience', really be derived from the success of our forbears in the pursuit of panthers? Monod does not explain what is supposed to correspond to those creative anticipations of nature in the survival skills developed by our remote ancestors. Do not our theories of life and the cosmos go beyond the attempt to describe 'events in the world around us, and to foresee their course'? What we wish to understand, *inter alia*, is how this world came about and developed into its present state. In pursuit of this aim we make theories which contain many assumptions about unobservable events, and which are connected to experience by means of inferences, the reliability of which would seem to be tenuously related to the practical requirements of survival in Cro-Magnon times. If man's survival were dependent on his power to formulate true theories about the origin of life and the cosmos, he would be as dead as the dodo.

We understand our mental and cognitive processes too faintly to place much confidence in any particular theory about how they developed. Here at least one may find some common ground with Monod. He says that we have no idea of the structure of what he calls 'the simulator' or of its functioning. How then can we hope to understand its origin and evolution? Uncertainty breeds uncertainty, and present doubts invade the past. So long as men's mental processes remain partly mysterious to themselves, this must surely affect any attempt to explain the manner of their genesis. As to whether these obscurities will ever be fully resolved, I prefer to remain agnostic, tolerant of an uncertainty I can do little to dispel.

I agree with Monod that the understanding of our mental processes has not advanced very greatly since Descartes. I do not understand how he can reconcile this with his very confident account of their evolution. Presumably he would agree that an adequate account of the origin of life could not, in principle, be formulated until the essential chemical structure of living things was understood. Belief in the intelligibility of a past cut off from all contemporary incomprehension, seems illogical. Any obscurities in our present condition come from the past. Life and mind do not mysteriously become more comprehensible as we follow them back in time. There is no Golden Age of ideal intelligibility and transparency awaiting us there, or haven from present perplexities.

I believe that the cognitive powers and interests we have inherited are deeper and wider than can confidently be accounted for on Darwinian lines, though possibly they are not so great as Darwinians assume them to be when they claim to understand how these powers developed. Man's desire to comprehend the universe, and to comprehend the comprehender, is absolutely without precedent in the history of life on earth. One should never underestimate the supreme audacity of the human animal in its aim to become, so to speak, an *anima Mundi* or world-intelligence. I cannot believe that this unique ambition, or the powers that prompt it, derive ultimately from random molecular changes, however the effects of these may have been selected. I believe such an explanation borders on being a pseudo-explanation, a way of concealing our essential ignorance from ourselves. 'Chance' is a marvellous cover-up word. Its indiscriminate use annuls one of man's finest traits, his awareness of ignorance. Only man knows that he is ignorant. He senses that his knowledge is partial and may always be so, applying to himself an ideal standard that may be unattainable and superhuman.

'To comprehend the comprehender': a tall order. How could one hope to comprehend or explain the value that men attach to right understanding, seeing that this valuation is presupposed by all attempts at understanding, including reflexive or self-oriented understanding? The belief that this value attached to right understanding is a by-product of its survival value is, as far as I can see, neither verifiable empirically nor especially plausible. Why *should* men's cognitive and contemplative values be derived from their practical needs? In the case of particular individuals, this is implausible. Why then should the derivation be accepted for the race in its evolutionary history? Because it brings man's quest for understanding into line with the prevailing biological conception of man and his evolution? But Biology itself depends on the value attached to understanding, irrespective of possible utility (or disutility). I am sure Darwin never thought that the value of his theories depended on their future possible applications. Darwin's dedication hardly strengthens Darwinism!

In its pursuit of understanding, science is essentially non-materialistic, unconcerned with material advantage or utility. It is questionable whether this scientific goal or value is compatible with its own child and product, in the case of Darwinism. If the latter were to induce scepticism about the capacity of men to labour in a disinterested way for the advancement of certain ideal aims, science would be the first to suffer, victim of its own teaching. So far as the scientist is inspired by the love of knowledge, he appears to live outside the domain of scientific materialism, and only to figure in this as an organism among other organisms when he is attending to his basic biological needs.

The world-view of science seems unable to comprehend the viewer. The scientist remains outside his theoretical constructions, transcending them, as their only begetter, and their continuous critic, arbiter, and reviser.

I shall now resume the discussion of certain points concerning the relation between science and religion, which have been prompted by the contents of Monod's book.

The view that religion can be invalidated by laboratory work seems to be on a par with the alleged remark of a Soviet astronaut that he had caught no glimpse of God in outer space. (Possibly he was not being ironical.)

Lucretius would not have laughed. He writes of divine beings that dwell in the Intermundia, or Between-Worlds. These live serene lives and are utterly indifferent to the affairs of men, having had no part in the creation of the world, and being themselves composed of the finest atoms, and having very thin bodies. Atomism will account even for the Gods.

Monod, if I have not misunderstood him, believes that Atomism will account for belief in the Gods, and their invention. For he maintains that there is probably a genetic basis for religion, in the innate categories of the brain, which have evolved selectively in the interests of social cohesion; and that this innate basis accounts for the fact, or alleged fact, that 'throughout the immense variety of our myths, our religions and philosophical ideologies, the same essential "form" always recurs'.[1] A rash generalization, surely. What it implies, in the religious domain, is that the Gods derive from certain formal myth-prescribing mutations in DNA at an early stage of human evolution. This is a neat reversal of the view that God made DNA. Since we are speaking of myths, it seems not unfair to describe this as a piece of molecular mythology, in all probability. To seek a molecular explanation of everything is the unique *déformation professionnelle* of the molecular biologist. This is not so much Neo-Darwinism as Neo-Lucretianism, or if one prefers, Democritean Darwinism. A molecule that gives not merely genetic instructions but also religious ones is a prodigy that even Lucretius might have had doubts about.

It would be equally possible in principle to think of Atomism, in bare analytical outline, as having been prescribed by genetic mutation. I find this as hard to believe as that random molecular convulsions should have prescribed the form of religious and philosophical ideologies. The first would imply that the DNA molecule, through random mutation, could initiate a way of thinking that would eventually culminate in a comprehensive theory of itself. I find this hard to believe, or indeed understand. At the very least, it suggests a new problem for geneticists, or moleculists: now they know how the DNA molecule copies itself, they might turn to the question how it was able to facilitate its own discovery and specification; its theoretical 'copying' or modelling. This problem might prove even more taxing; perhaps insoluble. At any rate, research in 'molecular epistemology' has not got started yet. It seems about as unpromising a subject as molecular theology, or molecular æsthetics or ethics.

Staggering creative powers are now ascribed, without a qualm, to the mutability of the DNA molecule. These powers are, to a large extent, inferred powers; their ascription is part of the great edifice of theory whose experimental vindication is confined to the sub-specific mutations whose effects have been experimentally studied. A giant inferential leap is taken, to the assertion that the inexhaustible richness of life is the product of the innate mutability of this supermolecule, aided by selection.

[1] *Chance and Necessity*, p. 156.

Never was so much owed to so little. In thus elevating one molecule above all others, Biochemistry has, as it were, reached the monotheistic stage of its development. This molecule is endowed, like God, with potential creative powers still greater than those exhibited in the realized world and the actualized evolutionary sequence; an implication of Neo-Darwinism is that in different environmental conditions, provided these were compatible with the emergence and continuation of life, a different range of fauna and flora would have arisen (by selection) out of the same creative cornucopia, the random mutability of DNA. The latter permits an unspecifiable variety of possible evolutions, of which the actually realized one is the one best adapted to the actually encountered conditions of life.

Not only is this immense first-order creative power ascribed to DNA (plus selection), but also the supreme second-order power of creating a being that is capable of creative activity. At this point surely it ought to have become clear that DNA has become invested with transcendental powers. *Expellas divum furca, tamen usque recurret.*

The model of DNA that has become so familiar a sight on all the 'media' has acquired the status almost of a scientific ikon, representing the creative source of life. And anyone who dares doubt its creative primacy is acting in a provocative, iconoclastic manner.[1]

Even on materialist assumptions, it is paradoxical to ascribe such vast innovative powers to a molecule, powers that greatly surpass those of the most complex known physical system, the human brain (itself a DNA-invention!). This may be why some physicists are inclined to be sceptical.

Personally, I find it inconceivable that a molecule, through its innate mutability, aided by selection, could have invented such a being as man, in his versatile innate potentialities. Far more conceivable is it that man should have 'invented' such a molecule —i.e., endowed it with creative powers that are, in part at least, imaginary and have certainly not been experimentally demonstrated. The history of human thought shows us other examples of creative power being attributed to physical entities that have

---

[1] If we believe that DNA is the ultimate creative source of all human and non-human life, past, present, and future, should we not make obeisance to it or its image?

later become discredited as sources and repositories. I very much doubt whether man's search for the creative source has now been successfully terminated, in a unique feat of enlightenment. It is terribly easy to exaggerate the long-term significance of contemporary discoveries (both in the sciences and the arts.) They have not been time-tested. We would be fools to believe that our age is uniquely immune from this error of self-aggrandizement. We should heed the rare wisdom of Einstein.

It seems to me that there is a real danger in the rapidity of scientific advance and the unremitting chase after the new. This leaves little time or opportunity for reflective assimilation of what has been achieved, and what has not, in the widest possible context. Like other *avant-gardes*, the scientific one suffers from a certain lack of historical perspective. He who looks back may get lost in the hunt.

A recurring human dream has been to reproduce biotic creation in the laboratory. In the second part of Goethe's *Faust*, Homunculus is manufactured in a test tube. Today's chemists would like to prove that their laboratory results hold the key to the understanding of past biotic creation or evolution. They seem often to underestimate the hazardousness of applying these results to the unique and so far unrepeatable evolutionary sequence culminating in the appearance of man. It seems possible that, by the nature of their work and the demands of their training, they are insensitive to the peculiar difficulties and frustrations of historical explanation—difficulties that usually become greater the more remote the events to be explained. Every competent historian is alive to the danger of converting the past into some kind of remote replica of the present. If he is honest, he will freely admit that there are some historical transformations which probably will never be adequately intelligible, for a variety of possible reasons. A belief held in common by Marxists and Darwinists, one of a Hegelian tint, is that they hold the explanatory formula which in principle embraces all social and organic transformations, however apparently wild and irregular.

I believe it to be historically irresponsible to assert categorically, on the basis of work in laboratories, that man originated through molecular mutation and selection. It seems to me that scarcely anyone would dare to assert this with any confidence, without the support of a philosophical faith and the solidarity which this engenders when widely shared. On the assumption that there must be a physical explanation, preferably of the atomistic type, for the evolution of man, molecular Darwinism seems the only plausible account. In any theoretical activity some assumptions are unavoidable, but we should not let their provisional status be wholly lost to view. We should not become their slaves and lose all capacity for criticizing them. A consensus may be comforting, but it may lead to complacency. The most dangerous assumptions are those which become truisms, which only cranks oppose—or 'idealists'. Even in pure mathematics, untouched by wordly perturbations and empirical discoveries, the most solid and long-standing consensus may eventually break down, and crucial hidden assumptions be exposed. The thoroughgoing evolutionist should be the last man to seek to freeze theories about evolution. Loyalty to 'the scientific world-view' is not an intellectual virtue. It can never be virtuous to convert assumptions into axioms.

The work of men does not easily, if ever, attain perfection, and limiting flaws appear, days or centuries later, that were invisible at the time of origin and maybe universal acclaim. Even the belief, now so confidently held, that forms the essential basis of molecular Darwinism—the belief that all hereditary characters are transmitted by DNA—may call for future modifications or qualification. Already it is possible to discern signs of this. One such sign has been mentioned in an earlier chapter. Men have an innately based sense of their own uniqueness and distinctiveness, their 'Iness', which cannot be simply derived from their unique genetic constitution, since identical twins are I's to themselves. The status of being an individual person, capable of being 'I' to himself, is not fully explicable in genetic terms.

A very different example of a possible limitation may be given. This is connected with a well-known problem concerning embryonic development, one of the deepest problems for any theory of development. The problem is to understand how in the course of development cells differentiate and form different tissues and organs, in an orderly manner or 'programme', even though the DNA in all cell nuclei is the same. In complete metamorphosis in insects, the discarded juvenile organs are broken down during pupation and the adult organs are developed. The same DNA specifies two distinct blueprints or body-plans, adapted to different ways of life, so that the young and the adult seem to belong to different classes, as phenotypically they do, though they are genetically identical. This encoding of two distinct organic plans in a single gene-complex is a remarkable phenomenon, and certainly a miracle (in the etymological sense).

DNA alone will not account for the programmed order of development, but DNA plus whatever controls and coordinates the activity of particular genes in different cells at different stages of development. Whatever the nature of this control may be, it could hardly itself be regarded as wholly under DNA control, by some feed-back arrangement. This control system must be an innate one, not environmental. From this it appears to follow that not all innate characters or functions are DNA-controlled or have evolved through selection of the effects of DNA-mutations.

The desire for unified explanations is very strong, and may be prejudicial. There is no *a priori* guarantee that a single explanation will in the long run prove adequate for all the varied phenomena of heredity, or organic evolution, or memory, or tradition, or cultural evolution. But unless we remain alert to the possibility of exceptions, actively seeking them, the eventual discovery of limitations to the established theory, if they exist, will be delayed. The right time to start looking for them is at the moment of triumph. All the philosopher can do is to question—and sow doubt. But this is not nothing. We all suffer from illusory certainties, and need to be undeceived.

A general limitation in present-day thinking about genetics derives from the experimental preoccupation with comparatively mindless forms of life and micro-organisms. This makes it easier to believe that heredity can be fully and adequately analysed without reference to 'mentalistic' concepts like purpose, the bugbear of biology. Now man, at least, inherits purposive and creative powers, exercised through partial recollection of the past and prevision, which need drawing out by education but cannot be created by it. Human foresight and purposiveness are innately based. How then can genetical processes be fully analysed and understood without reference to the concept of purpose? In the orthodox language of genetics, purposiveness is the functional expression of human genes under certain environmental conditions. How purposiveness can be the expression of the non-purposive is a great mystery. To understand how purposiveness can be inherited, we probably need a theory that has yet to be invented. Molecular biology is unpromising here. Even if a complete description could be given of all the chemical reactions which go on during the development of a human being from a fertilized egg-cell, the question would remain: how do these reactions make purpose and prevision possible? How can a purposive brain be the end-product of non-purposive chemical processes? The transformation of an egg-cell into a being capable of living in the future out of the past is so extreme a change as to make current conceptions of causality and development seem terribly thin. The only change that appears remotely comparable is the evolution of prevision and consciousness from primitive forms of life. It is orthodox today to say that the latter process was largely accidental and unpredictable, whereas the development of purposive capacities during ontogeny is the fulfilment of man's genetic destiny. This, however, does not appear to make the latter process far more comprehensible, when the fully developed capacities seem to surpass so greatly the processes of their development.

Like the neurophysiologist, the geneticist has to believe that the system he studies, with partial insight, is linked to purposive and creative activity, without being in a position to explain this linkage. A further point of similarity may be that for both inquiries, a question that arises is how a spatialized structure is translatable into a sequential programme. As the future development of the adult organism from the fertilized egg is 'hidden' or 'anticipated' in the molecular structure of its DNA, so it is possible to assume that when, for instance, a pianist performs the Hammerklavier Sonata from memory, his future play as he sits down is pre-represented, in its main outline, in his prepared brain state. Is it possible that this similarity might provide a conceptual link between the purposive behaviour of the adult and the programme of early development? Could there be a convergence on this question, how a future timed 'project' can be non-temporally represented by some equivalent form of order that pre-exists its temporal translation? This is a very tentative suggestion. The concept of purpose is very elusive, and its relationship to the processes of early development and genetics, as these are now conceived, may be such as to preclude complete clarity or the

hope of an early resolution. If the capacity for purposive behaviour depends in part on genetic inheritance, then it is reasonable
to say that the processes of heredity are purpose-oriented, even if
we are unable today or even tomorrow to analyse or comprehend
this orientation. The concept of purpose can only be banished
from biology if it is banished also from psychology, and this
would deprive both psychologists and biologists of any purpose. If
purposive concepts were retained in a Psychology cut off from
its biological base, this would imply an extreme mind-body split.

Einstein, whom I regard as the greatest natural philosopher of
our epoch, and who was capable of doubting the long-term
significance of some of his own achievements, would probably
have agreed with Goethe: only in limitation is mastery revealed.
The search for unlimited, all-embracing theories of life confronts
us with mysteries that may, as Einstein says, prove impenetrable
to human reason—however unfashionable this may sound. Many
such all-embracing systems have come and gone in the history of
human thought—though particular insights contained in them
may survive. Some of these systems have been developed by the
use of certain key analogies or metaphors, stretched and universa-
lized to create a single thought-world. One such instance is the
metaphor of the mirror for Leibniz, and another one is the
selection-metaphor for Darwin. Such unifying metaphors make
a strong imaginative and poetic appeal. They keep dryness away.
A recent example is that of the genetic code. Certain metaphors
may sum up a system, as certain images may characterize a
Shakespeare play. The importance of metaphors to human thought
and language, in all their domains, can hardly be exaggerated. In a
slightly metaphorical sense of 'metaphor', one may speak even
of visual metaphors in Art. Also possibly in some forms of music
in a rather more metaphorical sense. Wagner's leitmotivs are
developed and transferred in the course of the drama from their
original use to associated or analogous ones. Wagner, of course,
was an exceptionally 'literary' composer.

Any definition of 'mind' would be inadequate which left out
its metaphorical fecundity and play, a main source of poetic and
productive power, a true 'making'. Would it be wholly fanciful
to see in this inexhaustible fertility a possible clue to the evolution
of mental processes and their relationship to organic evolution as

a whole? A metaphor is a kind of incipient or potential metamorphosis, an ideal transformation that may or may not be 'actually' feasible. Men were called 'apes', abusively, long before Darwin. Such metaphors anticipate a full theory of metamorphosis.

The strong anti-metaphorical statement that everything is what it is, and not another thing, belongs in spirit to an anti-evolutionary view of the world. In its love of metaphor, the human mind mirrors the transforming processes of nature!—and is thereby predisposed to believe in the 'magical' possibility of everything being transformable? We live in thickets of metaphors, as Baudelaire nearly said. A Kantian might say that metaphorizing was an inborn tendency for the interpretation of experience. Strip thought of its metaphors and it would be bare indeed. Early immersion in poetry may be the best way of developing and strengthening this native productiveness of mind.

When some great change comes about that defies augury and analogy, we do not know what to think or how to think. There is good reason to believe that this planet was for a long, long time 'value-free', that its denizens were incapable of discriminating good from bad. Here is a signal and unique transformation that eludes human understanding.

One of the highest human values is love. (I hesitate to say this is *the* highest, lest it be thought to imply that any moment not consumed in love is time lost.) How love originated does not seem to me a 'soluble problem'. The notion that it originated 'by chance' may be put on a par with the thought that a computer could have composed *Tristan und Isolde*. If one believes this one will believe anything.

> Love makes strong plea
> Against fortuity.

Once love has developed, it becomes one of the great anti-chance factors in life. Whatever it is that one loves evokes one's solicitude and care, and a reluctance to 'leave anything to chance' in matters concerning its welfare and development. This may even be overdone, leading to over-protectiveness. But it is only when one is wholly indifferent to someone or something that 'the automatism of events' is tolerated.

How is it conceivable that love could have been 'engineered' or 'built in' by an evolutionary process indifferent to the realized quality and values of living? I find this almost a comical idea, inviting the attention of a satirist or cartoonist. But it is only comical on condition that, seeing the incongruity or absurdity, one withholds or suspends belief. If believed in, it can be nightmarish to think that our highest values are accidental products of an insentient automatic process. Can anyone really believe this with equanimity? Can anyone really believe this at all, if he reflects deeply on it? A general theory may be plausible in the abstract, and become incredible only when particular cases are considered. The notion that true theories must be wounding and offensive to our deepest feelings is absurd.[1] Turning Leibniz on his head creates more myths. It is not life that is wholly 'absurd' but our saying this.

There is always a personal element in one's responses to the enigmas of life. Personally, when confronted by the question how love originated, I get an acute sense of aporia not unmixed with awe. If, like Wagner's Mime, I had the choice of three questions to put to a superior intelligence, this would certainly be one of them. (However, the moral of the Wanderer-Mime questioning match is that it needs a superior intelligence to know which questions are truly important and revelatory.)

When one is out of one's depth it is natural to make for the shallows. The question how love originated may be thrown out; for instance, by saying that 'love' does not 'exist' any more than 'Time' does. Only events exist, only loving occurs. But this does not dispose of the question how loving originated.

There are other ways of disposing of the question, most of them behaviouristic and reductionist. I do not think they are adequate. The goal of reductionism is to deflate a question down to our present level of understanding, and so to reduce the peaks of human experience. It is a kind of theoretical debunking, or devaluation. Occasionally it may even be justified. But it is dangerous, as its effects may spread beyond the theoretical or academic level. Theory and practice intertwine.[2]

A world without love would be as intolerable for beings like ourselves as a world without water.

---

[1] The 'Dissonance' theory of truth.
[2] Pascal's *Pensées* include this one: 'It is dangerous to explain too clearly to man how like he is to the animals without pointing out his greatness.'

How a world without love could have generated love, I do not
aspire to know. I do not think any man knows this.

This seems to me a far greater miracle than anything officially
catalogued as such.

Index